The Physics of Evolution

This book provides an introduction to the significant role of physics in evolution, based on the ideas of matter and energy resource flow, organism self-copying, and ecological change. The text employs these ideas to create quantitative models for important evolutionary processes.

Many fields of science and engineering have come up against the problem of complex design—when details become so numerous that computer power alone cannot make progress. Nature solved the complex-design problem using evolution, yet how it did so have been a mystery. Both laboratory experiments and computer-simulation attempts eventually stopped evolving. Something more than Darwin's ideas of heredity, variation, and selection was needed.

The solution is that there is a fourth element to evolution: ecological change. When a new variation is selected, this can change the ecology, and the new ecology can create new opportunities for even more new variations to be selected. Through this endless cycle, complexity can grow automatically. This book uses the physics of resource flow to describe this process in detail, developing quantitative models for many evolutionary processes, including selection, multicellularity, coevolution, sexual reproduction, and the Serengeti Rules. The text demonstrates that these models are in conceptual agreement with numerous examples of biological phenomena, and reveals, through physics, how complex design can arise naturally.

This will serve as a key text on the part physics plays in evolution and will be of great interest to students at the university level and above studying biophysics, physics, systems biology, and related fields.

Author Biography

After getting a PhD in physics at the University of Illinois at Urbana-Champaign, Michael Roth was inspired by Richard Dawkins' *The Blind Watchmaker* and the study of evolution became a nearly lifelong interest for him. Following a postdoc at Fermilab, he joined The Johns Hopkins University Applied Physics Laboratory (JHUAPL) where he served in many scientific leadership roles until retirement. He has major publications in several fields including theoretical physics, geophysics, neural networks (a.k.a. deep learning), biophysics, and instrumentation. He was a co-chairman for several conferences. He has also served as a JHUAPL group supervisor and as the Book Review Editor and Editor-in-Chief of *IEEE Transactions on Neural Networks*. After retirement from JHUAPL, he devoted his time to his longtime interest in evolution.

The Physics of Evolution

Michael W. Roth

CRC Press
Taylor & Francis Group
Boca Raton London New York

CRC Press is an imprint of the
Taylor & Francis Group, an **informa** business

First edition published 2023
by CRC Press
6000 Broken Sound Parkway NW, Suite 300, Boca Raton, FL 33487–2742

and by CRC Press
4 Park Square, Milton Park, Abingdon, Oxon, OX14 4RN

CRC Press is an imprint of Taylor & Francis Group, LLC

© 2023 Michael W. Roth

ISBN: 9781032489070 (hbk)
ISBN: 9781032490427 (pbk)
ISBN: 9781003391395 (ebk)

DOI: 10.1201/9781003391395

Typeset in Minion
by Newgen Publishing UK

Contents

Preface

ONE MIGHT WONDER WHY physics has not made more connections with biology. Several authors have proposed that physics and biology are strongly linked. For example, many of the tools used for biological research (e.g., X-ray crystallography for DNA structure and MRI for many uses) were developed by physicists. There are excellent textbooks on biophysics (e.g., Bialek, 2012 and Forgacs and Newman, 2005). The work in systems biology has been strongly influenced by physicists (e.g., Alon, 2006). The importance of physics constraints and their influence on evolution has been noted (Cockell, 2018). And finally, biophysical rules can provide a basis for the emergence of the genetic code (Harrison et al., 2022).

However, chemistry and engineering have made stronger connections with biology, so why not physics? Biophysics has been a recognized academic subject for many years, but it has not yet attained the visibility and interest of other cross-field subjects like biochemistry and bioengineering. One could speculate why this is true. Perhaps it is partly due to the emphasis in physics on mathematical models. But biology now heavily employs mathematical models from DNA matching to sophisticated statistical data analysis to ecological simulations for climate change. Perhaps it is partly due to the generally reductionist approach of physics that tries to make the simplest models that still represent the data, whereas many biologists are attracted to the sheer complexity of biological systems.

I propose that closer connections between physics and biology could be more numerous if physics could contribute more to the theory of evolution. Evolution theory is central to biology and nothing in it makes sense otherwise (Dobzhansky, 1973). Biological organisms cannot exist without the flow of energy-and-matter resources. Without such flow, reproduction and metabolism stop and species go extinct. So energy-and-matter-resource flow must be central to evolution. But energy-and-matter flow has long been a major topic in physics from thermodynamics to fluid dynamics to electricity and beyond. So could any insights to evolution be gained from the perspective of the physical flow of energy-and-matter resources? The purpose of this book is to explore that question in more detail.

The physics models of energy-and-matter flow differ from case to case, but there are common features of interest such as conservation and limits. This is also true for models of biological systems. So we will head down this path of defining models for biology and evolution and see where that leads. We will start with the simplest cases corresponding to the origin of life and gradually increase the complexities of the environments and systems. At each stage, we will reflect what we have observed and employ that knowledge for the next stage. There are certain critical factors for the emergence of complex systems and we will pay particularly close attention to them. In the end, we will show that physics does indeed have much to say about the evolution of complex biological systems.

There are frequently modern analogs of species that arose over the course of evolutionary time. For instance, there are a large number of modern marine animals with different kinds of vision systems from light-sensor patches to complex eyes. When lined up, it becomes more obvious that there was a progressive evolution in the development of the complex eye. We shall make extensive use of the observations of these kinds of modern analogs.

Richard Dawkins (1976) postulated that evolution could be understood by the existence of things that made copies of themselves, which he called replicators. In addition, Sean Carroll (2016) postulated a series of rules for regulating ecologies that he called the Serengeti Rules. He was especially interested in ecological change and the rules were highly relevant to that. We shall make extensive use of these ideas. We shall end up with a series of models that reflect stages of evolution and ecological change.

But at the heart of this book is the desire to learn how complexity and novelty can arise automatically. Many fields of science and engineering have come against the problem of complex design. This occurs when the details become so numerous that computer power alone cannot make progress. For neural networks (a.k.a. deep learning), this is known as the NP-completeness problem whereby the amount of computation needed to compute network weights grows in a non-polynomial (e.g., exponential) way as the number of weights increases. For systems biology, it happens when there are too many cell-model parameters to allow for successful estimation. For many engineering fields, it happens when the designs become so complex that integrated testing cannot eliminate faults.

Nature solved this complex-design problem using evolution. But exactly how it solved it has been a mystery. Both laboratory experiments and computer-simulation attempts eventually stopped evolving. Something more than Darwin's ideas of heredity, variation, and selection was needed. The solution is that there is a fourth element to evolution and that is ecological change. What happens is that a new variation that is selected can change the ecology and the new ecology can create new opportunities for even more new variations to be selected. By this endless cycle, complexity can grow automatically. This book uses the physics of resource flow to describe in detail how that happens and that knowledge can be used to help overcome the problem of complex design.

This book is a monograph that can be used as a textbook on how physics plays a strong role in evolution. The key starting ideas are the flow of energy-and-matter resources, organisms making copies of themselves, and ecological change. From these ideas, quantitative models are developed for many evolutionary processes. Such models include selection, multicellularity, coevolution, sexual reproduction, and the Serengeti Rules. Numerous examples of biological phenomena are shown to be in conceptual agreement with the models. In the end, the physics shows how complex design can arise automatically.

A summary of the book chapters is as follows. Chapter 1: Fundamentals. Current evolutionary theory is reviewed and the need to consider ecological change is discussed. The method of walking through evolutionary history is established and started. The leading theory of the origin of life is discussed and the basics of the models are described. The importance of detritus and its current manifestation as marine snow is also discussed. Then, the central idea of the Selection Theorem is proven which gives a key quantitative meaning to natural selection. In addition, it is shown that resource preference leads to coexistence which is the foundation for cooperation and symbiosis.

Chapter 2: Fertility. Initially, the mortality of offspring must have been considerable. Like many things in evolution, there are modern analogs so the case of human juvenile mortality is discussed as an example and variations with less mortality are selected. Invasive species with greater fertility can result in selection and an example is the Great American Interchange that occurred with the forming of the Isthmus of Panama. But once metabolism and the genetic code are established, evolution can delete useless machinery in the phenomenon known as streamlining. Early multicellular evolution is discussed, and critical factors are identified. Finally, the hazard to evolvability from excess fertility is discussed. Fertility has the power to build, replace, and simplify as well as curb evolution.

Chapter 3: Scavengers. With a planet full of detritus available, replicator variations appear that can consume various ages of detritus. Specialists for particular consumption and generalists for variety consumption appear and compete. The result is that generalists have the advantage unless the specialists are very fertile. Generalists also compete, but unless the resources and resource preferences are identical, the generalists will coexist. In simpler environments, replicators will evolve towards simpler forms; while in more complex environments, replicators will evolve towards more complex forms. Evolution is about adaptation to whatever environments are present.

Chapter 4: Predators. At some point, it appeared to be an easy evolutionary step to go from consuming fresh detritus to consuming live prey. But this is actually a remarkably difficult step because of the danger of extinction from overfeeding and cannibalism. Predator generalization helps, but in some sense just delays the inevitable. It was the evolution of prey defense that helped prevent predator extinction from overfeeding. And kin selection had to evolve to create an aversion to cannibalism. Finally, there were predator-prey cycles that came dangerously close to extinction. The models reflect all of these phenomena. When these challenges are met, the stage is set for an explosion in species diversity.

Chapter 5: Arms Races. Fertility and scavenging were enough to build metabolism and genetic codes, but the great leap to complex multicellular animals took a big push. And that was predator-prey arms races. This chapter describes the models of arms races and how novelty arises automatically. While prey were under some selection pressure, predators were under stronger selection pressure to evolve capabilities to overcome prey countermeasures. Eventually, an arms race could become a Red Queen's scenario where predators and prey were evolving as fast as possible but appeared as if little progress was made. Under this scenario, a myriad of different predator-offensive and prey-defensive traits would grow automatically depending on the evolutionary trajectory. Once again, biological phenomena are shown to agree with the models.

Chapter 6: Trophic Cascades. At this point, the models have described the rise in complex animals with novel characteristics. But what about complex ecologies? In this chapter, we take the first step in showing that the models can also describe complex ecologies and their changes. We start with examples of trophic cascades whereby changes in the top of a food web can have large effects throughout all lower levels of the web.

The cases of the orca-otter-urchin-kelp and starfish-mussel-other trophic cascades are discussed. Again, the models can successfully describe these phenomena. In addition, the reader learns that there are evolutionary consequences from trophic cascades because highly suppressed species go through a founder effect whereby the survivors have greatly reduced genetic diversity.

Chapter 7: Parasites and Pathogens. Nowhere is evolution more manifest than with infectious diseases from parasites and pathogens. And it represents another example of complex ecologies. The case study here is the use of the MYX virus against the Australian rabbit overpopulation. This case shows all of the stages of a strong pandemic: virulence, attenuation, resistance, and coevolution. The models are consistent with all of these stages. Because of potentially high mortality, arms races with infectious diseases can create very strong selection pressure for disease resistance and better immune systems for hosts. This is the driver for the evolution of sexual reproduction. A detailed model is presented that shows that sexual species can gradually evolve from asexual species under specific evolutionary trajectories.

Chapter 8: Serengeti. The previous three chapters showed how arms races could cause the emergence of ever more complex and novel species and ecologies and that the models can describe them. This chapter now shows how the elements of the models can be combined to describe the upgrading event for the complex Serengeti ecosystem. After a few more model elements involving migration, residency, grass competition, and wildfires are developed, this chapter describes the Serengeti ecosystem model derived from the models. When the Serengeti rinderpest pandemic was eliminated by vaccination, the wildebeest population increased fivefold causing an upgrading trophic cascade for the ecosystem and the model shows conceptual agreement. Sean Carroll (2016) developed a set of ecological rules to help understand trophic cascades and the Serengeti upgrading which he called the Serengeti Rules. The models developed in this book are a representation of the Serengeti Rules.

Chapter 9: Summary Discussion. The book ends with a discussion of the key ideas presented in previous chapters along with recognition of select members of the scientific community for their critical contributions.

Appendix A: Density-Dependent Regulation. The models use threshold regulation rather than conventional ecology's logistic regulation for density dependence. This technical appendix discusses the challenges of estimating density dependence from biological data.

Appendix B: Selection Theorem Proofs. This technical appendix contains the complete proof of the Selection Theorem.

The author would like to thank Richard Dawkins, Harold Morowitz, Nick Lane, Sean Carroll, Nigel Goldenfeld, Charles Cockell, Wallace Arthur, and Laura Portwood-Stacer for their ideas. The author also thanks his wife, Susan, for numerous discussions and her comments.

CHAPTER 1

Fundamentals

FROM THE ZOO OF elementary particles to the unfolding of the cosmos, physics is all around us. Physics provides the knowledge of form and function, basics and possibilities. Although physics may seem disjoint from biology, it is not. In fact, it is central. Everything in biology has important physical aspects without which biology would not be possible. From the mechanisms of locomotion to the pressure from diaphragms that allow human breath, physics is at work. All biological organisms exist because of the flow of energy and matter. It is precisely this flow that brings essential resources so that organisms can function, grow, and reproduce. But the flow of energy and matter is a physical phenomenon for which physics has much to say.

But evolution is the most important phenomenon in the universe. It changes dead and poisonous worlds into green planets with occupants that can ask why. Evolution is at the heart of understanding all biological organisms. As Dobzhansky (1973) said, "Nothing in biology makes sense except in the light of evolution." Clearly, chemistry has played a strong role for biology and evolution in providing building blocks for structure, function, and hereditary information.

But what is the role of physics in evolution? Life exists because of the flow of energy and matter. Without this flow, the resources necessary for

DOI: 10.1201/9781003391395-1

the creation and maintenance of life would not be there. The flow of energy and matter is a subject of considerable study in physics. For the dynamics of fluids, gases, and plasmas, physics provides insight and knowledge. So could the physics of flow provide additional insight to understanding more about evolution? That is the subject of this book which will show that physics does indeed have much to say about evolution.

It is a fact that biological history is a consequence of evolution. Comparative genomics proves this beyond any rational doubt (Dawkins, 2009). The resulting great and sometimes tangled tree of life is due to evolution. The "what," as in "what happened," for evolution is on very solid scientific ground from the Last Universal Common Ancestor (LUCA) onwards.

But the "why" of evolutionary theory continues to evolve. Darwin's theory of evolution relied on the central elements of heredity, variation, and selection (Darwin, 1859). But Darwin's original model of heredity by blending inheritance was incomplete and many realized that his idea of natural selection didn't work with it. Consequently, his ideas took many years to be accepted among the biological-science community (Ridley, 2004, pp. 12–18 and 37–41). It was Mendel who realized that heredity did not blend but instead was quantized (Mendel, 1866). The importance of Mendel's work was not initially realized until its rediscovery (Bateson, 1909) changed everything.

It was the mathematical models of Fisher (1918, 1930), Haldane (1924), and Wright (1931) that showed that a Mendelian model of heredity fixed Darwin's theory and made it work by postulating the quantum of heredity known as "genes." And it was the publication of Huxley's book (Huxley, 1942) describing these models that enlightened the larger biological-science community to these ideas. It was only then after World War II that the revised theory of evolution became widely accepted by biologists. Evolution theory made more sense, but the need to grow continued.

In the 1940s and early 1950s, experiments pointed to DNA as the molecule responsible for biological heredity. The discovery of the chemical structure of DNA (Watson and Crick, 1953) marked the beginning of a long arc of discoveries culminating in genomics. At first, it was thought that the protein-coding sections of DNA were the genes of heredity (Gericke and Hagberg, 2007). But then, it was realized that there was much more to heredity because many species with vastly different characteristics had about the same number of protein-coding sections. The solution was the realization that regulatory sections of DNA were also genes. For example, during development regulatory sections known as switches turn protein-coding sections on and off (Carroll, 2008). It is the regulatory sections together with the protein-coding sections that are the quanta of heredity.

Even with all this scientific achievement, many have expressed the need to update the evolutionary theory described by Huxley. A sampling of the ideas offered include, but are not limited to, selfish genes (Dawkins, 1976), endosymbiosis (Sagan (née Margulis), 1967; Margulis, 1970), and the consequent energetics (Lane, 2015), neutral theory (Kimura, 1968 and 1991), game theory (Maynard-Smith, 1979), punctuated equilibrium (Eldredge and Gould, 1972), evolutionary innovations (Wagner, 2014), and the need to merge with genomics (Pigliucci and Mueller, 2010). Mathematical treatment for some of the concepts can be found in Rice (2004). Hopefully someday, consensus mathematical models will emerge that incorporate all the complexity of genomics into evolutionary theory and when that happens it will be a monumental scientific achievement.

But a major piece that is still missing is the influence of environmental change. Without it, both laboratory experiments and computer simulations (see, e.g., Bedau, et al., 2000; Standish, 2003; Packard, et al., 2019) of evolution eventually stop evolving. For example, Lenski's long-term multigenerational environmentally static *E. coli* experiment shows an ever slowing of average change (Wiser, et al., 2013). Dawkins (1986, pp. 178–179) describes the importance of environmental change as follows:

> Evolution will come to a standstill until something in the conditions change: the onset of an ice age, a change in the average rainfall of the area, a shift in the prevailing wind. ... The weather is very important to animals and plants. Its patterns change as the centuries go by, so this keeps evolution constantly in motion as it "tracks" the changes. But weather patterns change in a haphazard inconsistent way. There are other parts of an animal's environment that change in more consistently malevolent directions, and that also need to be "tracked." These parts of the environment are living things themselves... And, just as long-term fluctuations in the weather are tracked by evolution, so long-term changes in the habits or weaponry of predators will be tracked by evolutionary changes in their prey. And vice versa, of course.

Environmental change is composed of both geophysical change like weather shifts and ecological change like the appearance of fiercer predators.

But geophysical change is not very interactive. For example, weather can change evolution, but evolution doesn't very often change weather except at large scales such as the great oxidation event. However, ecological change is very interactive with species often appearing and causing the disappearance of other species. It is this ecological change that is of great interest because it is the interactive characteristic that has the most potential for building complexity and novelty.

But how can one scientifically investigate the effect of ecological change on evolution? Paleontology is not expected to be of much help in this because of the number of gaps in the fossil record. There have been some successful multi-year field studies of evolution and geophysical change. Examples include beak-size changes in Darwin's Galapagos finches due to the appearance of very dry or wet seasons (Grant and Grant, 2006) and the variation in melanism in peppered moths due to changes in industrial pollution (Ridley, 2004, pp. 108–114). There are now several examples of ecological changes with evolutionary implications due to trophic cascades (Estes et al, 2011). In addition, careful thought experiments have provided substantial insight into the effect of ecological change on evolution (Dawkins and Krebs, 1979). The importance of ecological-evolutionary feedback has been noted (Murugan et al., 2021). And an extensive review of much of the state of research can be found in Hendry (2017).

But more is needed in order to explore evolution and ecological change. For that we must turn once again to mathematical models. Nilsson and Pelger (1994) faced a similar dilemma when they wanted to explore the evolution of as complex an organ as the human eye. Their ingenious solution was to postulate reasonable evolutionary stages of development from a light-sensitive eye patch, to arrays of patches, to forming a cup of surrounding tissue, to a nearly pinhole camera, to a clear cover for the camera, to an initial lens formation, to the completed eye with a focusing lens. Interestingly enough, examples of almost all of these stages have been found in present-day biological organisms (Emlen and Zimmer, 2020, pp. 336–342). Nilsson and Pelger postulated that the optical physics measure of visual acuity was a reasonable measure of fitness. Then by a series of small variations, there was a reasonable and gradual series of evolutionary steps from beginning to end for the formation of the complex eye. When they combined their model with estimates of genetic mutation, they concluded that the evolution of the eye could occur in a surprisingly small number of generations.

We shall follow the general approach of Nilsson and Pelger toward building models of evolution and ecological change. Like them, we shall postulate stages of evolution and determine what we can learn. We shall use models of ecology that are related to conventional ecological theory (e.g., Soetaert and Herman, 2009). But there is a significant difference here. We shall strip the models down to their barest critical elements and employ those. In that vein, we shall diverge from Nilsson and Pelger and not postulate a fitness function to drive selection. Instead, we shall use the dynamics of ecology to determine survival. Note that ecological models are very much concerned with the flow of resources. And it is on the effects of changes in resource flow that we will concentrate.

This use of highly simplified models is a common approach in physics and our motivation is similar. We do not wish to follow the usual path in ecological modeling of creating detailed models for comparing to empirical data. In particular, there is not a lot of empirical data on evolution and ecological change to go around. Instead, we are interested in illuminating the processes of evolution and ecological change to see what we can learn. Highly simplified models can serve that purpose. As Albert Einstein once said (Einstein and Calaprice, 2011):

> It can scarcely be denied that the supreme goal of all theory is to make the irreducible basic elements as simple and as few as possible without having to surrender the adequate representation of a single datum of experience.

This is our goal: to illuminate the "why" of evolution and ecological change and learn how complexity and novelty arise automatically.

1.1 ORIGIN

Consider a world utterly devoid of life. The volcanoes and meteor bombardments have settled down some, but there is no breathable atmosphere, mostly carbon dioxide with no appreciable oxygen. But there is water. All the volcanic action has spewed up enough water vapor to condense and form oceans. Tectonic action has created oceanic ridges and hydrothermal vents.

Not far from the oceanic ridges, an unusual non-volcanic kind of vent appears, known as an alkaline hydrothermal vent. Alkaline vents are very different. They are created by water seeping down into the rock below the ocean floor, reacting with it, and spewing up warm alkaline fluids containing iron sulfide and dissolved hydrogen. The fluids precipitate to form chimneys with labyrinths of micropores. The fluids flow over catalytic surfaces containing the iron sulfide. The hydrogen reacts with the carbon dioxide in the water via catalytic micropore channels to form organic molecules (Martin and Russell, 2003; Hudson et al., 2020). Because the organics are heavier than other molecules, they begin to accumulate in the micropores and interact with each other. Thus, the alkaline vent begins to act as an electrochemical flow reactor. Lane (2015) describes it as follows, relative to the primordial-soup concept of the origin of life (Oparin, 1924; Haldane, 1929):

> In the absence of genes or information, certain cell structures, such as membranes and polypeptides, should form spontaneously, so long as there is a continuous supply of reactive precursors— activated amino acids, nucleotides, fatty acids; so long as there is a continuous flux of energy providing the requisite building blocks. Cell structures are forced into existence by the flux of energy and matter. The parts can be replaced but the structure is stable and will persist for as long as the flux persists. This continuous flux of energy and matter is precisely what is missing from the primordial soup. There is nothing in soup that can drive the formation of the dissipative structures that we call cells, nothing to make these cells grow and divide, and come alive, all in the absence of enzymes that channel and drive metabolism.

In the micropore labyrinths of the alkaline vents, long-chain polymers start to accumulate along with membrane lipids and nucleotides. The organics increase and flow past the organic-filled micropores until a single membrane bag of organics in a micropore is filled with so many organics that it splits into two bags. Thus, an entity has made a copy of itself. This is the first replicator.

Dawkins (1976) proposed the concept of replicators as a way to understand evolution. His idea was that once replicators appeared, then Darwinian evolution would automatically occur with heredity, variation, and selection. As shown in Figure 1.1, a replicator would consume resources, make copies of itself, and become detritus if resources were exhausted.

Variations in replicators would create competition and selection of ones with survival advantages. Dawkins' concept was qualitative in nature, but we shall show that quantitative models of replicators can illuminate many interesting aspects of evolution.

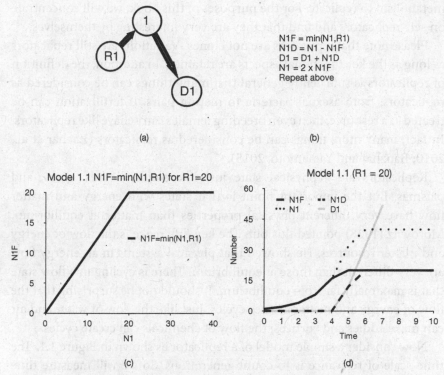

Model 1.1 N1F=min(N1,R1) for R1=20

Model 1.1 (R1 = 20)

(a)

(b)

(c)

(d)

FIGURE 1.1 Simple Replicator Model 1.1. (a) Food web. Replicator 1 consumes resource R1, make copies of itself, and becomes detritus D1 if the resource is exhausted. (b) Formulas. N1 is the population of replicator 1. R1 is the amount of resource R1 per unit time. N1F is the final population of replicator 1 after consumption and expiration. N1D is the amount of replicator 1 detritus created per unit time. D1 is the amount of accumulated detritus. After replication, the population N1 is twice the final population N1F and the cycle repeats. (c) Plot of N1F=min((N1,R1) for R1=20. (d) Results. R1=20. At t=0, N1=1.

We shall define replicators as follows:

> Definition: A replicator is a unit of energy-and-matter combination that can make a functional copy of itself.

By this definition, exact copying is not required, only copying key functional aspects such as the ability to replicate, consume vital resources, and expire without resources. This definition differs from Dawkins' in that we are focusing on self-replicators. Dawkins was interested in genes as replicators. But genes are not self-replicators because they rely on metabolism to replicate. Viruses are also not self-replicators because they hijack metabolism to replicate. For the purposes of this book, we will concentrate on self-replicators and find that they are very interesting by themselves.

Please note that replicators are not clones. Variations are still replicators as long as the key functional aspects are retained. In addition, the definition of replicators is sufficiently general that many things can be considered as replicators, from asexual bacteria to mating pairs. If fertilization can be treated as a resource, then even breeding females can behave like replicators. In fact, many more things can be considered as replicators (Zachar et al., 2010; Banzhaf and Yamamoto, 2015).

Replicators are a physical state just like solids, liquids, gases, and plasmas. But the important point is that states with energy and matter flow have very different physical properties than matter at equilibrium. Morowitz (1968) pointed this out. The key difference is the flow of energy and matter resources. He showed that physical systems in an energy flow are very different than those at equilibrium. There is cycling in a flow state that is maximally far from equilibrium. It should not be surprising that the flow of energy and matter creates cycles. Just like the flow of water and air can make eddies and vortices, the flow of chemicals can create cycles.

Now consider a simple model of a replicator as shown in Figure 1.1. The time scale of relevance is to count generations. So we will measure time in discrete units corresponding to the time needed to obtain the necessary resources and replicate. We will approximate resources by lumping them all together into a single resource. This is not expected to be a bad approximation because replication will always be limited by some critical resource. The limiting factor is exhaustion of the resource that would be the first to stop replication. In biology, this is known as Liebig's Law of the Minimum: in any ecosystem, there will be some limiting nutritional factor that will determine growth (Atlas and Barth, 1997, pp. 281–282). It is the flow of the limiting resource that determines the amount of replication. Finally, the resource scale of relevance is the amount needed to survive and replicate. So, we will measure resources in that unit. One unit of resource enables one replication.

Figure 1.1a shows the food web for this simple model. With the beginning of each time step, the flow of resources provides R1 units of resources. Replicator 1 then consumes the resources of one unit per member of the population. Replicators that succeed in consuming a resource unit then replicate. If there are too many replicators, resource R1 runs out. Replicators that do not consume a resource unit become detritus D1. Figure 1.1b shows the corresponding formulas.

Simple Replicator Model 1.1 Details. N1 is the population number of replicator 1. R1 is the amount of resources per unit time. If N1<R1, then the number fed N1F is equal to N1. But if N1>=R1, then the number fed is N1F=R1. The formula N1F=min(N1,R1) is just a convenient way to represent both cases and is plotted in Figure 1.1c. The amount of detritus created is N1D=N1-N1F and the accumulated detritus is D1=D1+N1D. The fed replicators get to replicate and the new population is now N1=2 N1F and the cycle repeats.

Figure 1.1d shows how the replicator population changes with time for R1=20 and starting populations of N1=1 and D1=0. At first, the population grows exponentially because there are plenty of resources to go around. But then, the population hits the limit of the resources and stops growing. The number fed at the resource limit N1F=R1 is called the carrying capacity. Now note that when the population hits the carrying capacity, detritus is created and accumulated detritus D1 grows by the carrying capacity with each generation.

So how does our simple model compare with more conventional ecological models?

Comparison Details. First, let's generalize replication to an arbitrary number of offspring. Let K be the replication ratio so that

$$N1 = K \, N1F \tag{1.1}$$

Now compute the new number of fed replicators in terms of the previous fed number. So

$$N1F = \min(N1, R1) = \min(K \, N1F, R1) \tag{1.2}$$

where the N1F on the left side of the equation is the new number and the one of the right side is the old number. Now compute the difference $\Delta N1F$ between the two:

$$\Delta N1F = \min(K \, N1F, R1) - N1F \tag{1.3}$$

This is a difference equation that is merely a discrete form of a differential equation. This replicator equation is not that different from conventional ecological theory. Conventional ecology theorists prefer more smoothly varying functions, which can be easier for solving differential equations, so they would rewrite the above difference equation using the logistic function (Soetaert and Herman, 2009, p. 276).

$$\Delta N1F = (K-1)\ N1F\ (1 - N1F/R1) \tag{1.4}$$

Figure 1.1e shows a comparison between the replicator-model $\Delta N1F$ and the logistic-function $\Delta N1F$ for K=2

This dependence of growth rate on population number is called density-dependent regulation. Both replicator and conventional-ecology logistic models have this characteristic. The growth rate is the same K-1 for small populations leading to the same initial exponential growth. The carrying capacity is the same R1. The growth rate is also the same near the carrying capacity for K=2. One difference is the different growth rate when the population is between zero and the carrying capacity. The replicator model regards the carrying capacity as a hard limit while the logistic function treats it as a softer limit. Per Einstein's dictum, we will stick with the hard-limit replicator model because of its simpler physical interpretation. However,

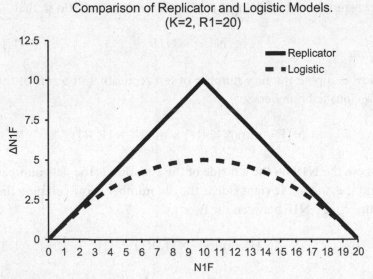

FIGURE 1.1E Comparison of density-dependent regulation for the replicator model Eq. 1.3 and the conventional-ecology-theory logistics model Eq. 1.4 with replication ratio K=2 and carrying capacity R1=20.

see Appendix A for a discussion of complications for measuring density-dependent regulation.

1.2 MARINE SNOW

It's snowing outside. You are sitting in your deep-submergence vehicle, or imagining it, on the bottom of the ocean. It's so dark outside that you had to turn on the outside lights to see anything outside of the view window. And it's snowing outside. How could it be snowing outside? It's called marine snow. Detritus from above clumps together and sinks to the bottom at such a rate that it looks like snow. Wrecked ships are covered in a few years. Over hundreds of years, such ships can be completely hidden. Over a million years, it can be 20 feet deep. Many animals and bacteria scavenge the marine snow on the bottom. There are plenty of organics in the snow to use as a primary food source (Silver, 2015).

William Beebe (1934) was the first to observe marine snow by using a bathysphere to descend to the ocean depths. Rachel Carson worked with Beebe early in her career and described it as follows (Carson, 1961):

> When I think of the floor of the deep sea, the single, overwhelming fact that possesses my imagination is the accumulation of sediments. I see always the steady, unremitting, downward drift of materials from above, flake upon flake, layer upon layer—a drift that has continued for hundreds of millions of years, that will go on as long as there are seas and continents. For the sediments are the materials of the most stupendous "snowfall" the earth has ever seen. … Silently, endlessly, with the deliberation of earth processes that can afford to be slow because they have so much time for completion, the accumulation of the sediments has proceeded. So little in a year, or in a human lifetime, but so enormous an amount in the life of earth and sea.

Before life started, there was no life-created marine snow. But when life did start, the ocean had the equivalent of ski-resort snowmaking guns: alkaline hydrothermal vents. Not only did the vents produce organics and living organisms but when the populations of life hit the carrying-capacity resource-limit the vents produced detritus as marine snow.

Over time, vast amounts of detritus can be created. If life originated from a hydrothermal vent, it is possible that with replicators created by a single hydrothermal flow, a generation cycle time of 30 minutes (typical of bacteria), natural water and air currents, and a billion years of elapsed time, the Earth could be covered in many layers of detritus. This detritus could literally fertilize the entire planet. Variations of replicators could use the detritus as a resource for consumption. With mixing, such as that from water and air currents, and negligible depletion, a detritus resource could be approximated as a depot and a flow.

Figure 1.2 shows Detritus Model 1.2. Detritus D2 is derived from the large time accumulation and distribution of detritus from replicator 1.

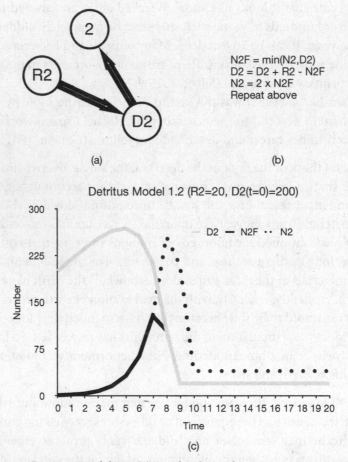

N2F = min(N2,D2)
D2 = D2 + R2 - N2F
N2 = 2 x N2F
Repeat above

(a)

(b)

(c)

FIGURE 1.2 Detritus Model 1.2. (a) Food web. Replicator 2 is a variation that can consume detritus which is modeled as a storage depot D2 with a resource inflow R2. (b) Formulas. N2 is the population of replicator 2 with N2F being the number fed. D2 is the amount of detritus available which is increased by resource inflow R2 and decreased by consumption. (c) Results. Replicator 2 population blooms until depot exhaustion causes a crash and eventual reliance on resource flow R2.

A variation of replicator 1 creates replicator 2 that can consume detritus D2.

> Detritus Model 1.2 Details. Detritus is modeled as a storage depot D2 with addition from resource flow R2. R2 models inflow of detritus from mixing currents to the replicator 2 area. Up to the available units of D2, each member of population 2 consumes one unit of D2 per unit time necessary for replication. N2F is the number fed. The population number N2 doubles N2F with each unit time cycle. Detritus production by replicator 2 is implicit.

Figure 1.2c shows the results of Detritus Model 1.2. The population N2 rises exponentially until the depot stock nears depletion. Then the depot stock and the population fall until the resource flow R2 is established as the carrying capacity. Note that it normally requires multiple units of Model 1.1 detritus D1 to be one unit of feed for Model 1.2 so they have different scales. Although detritus from replicator 2 is neglected because its effect is small, depletion of the depot D2 causes Model 1.2 to become essentially Model 1.1.

1.3 SELECTION

Lenski's long-term experiment on *E. coli* evolution has yielded numerous insights on the processes of biological evolution in a relatively simple model system (Lenski, 2017). By providing a flow of glucose-nutrient resource to a set of *E. coli* cultures, his team was able to observe and dissect evolutionary changes in the populations. We have already mentioned the observation of the expected slowdown in average change because of the static environment.

But one of the most important observations was a dramatic example of variation and selection. Citrate was added to the nutrient medium so that needed iron uptake could occur and the colonies would not perish prematurely. This was anticipated to have a relatively neutral consequence because *E. coli* were not expected to consume citrate in the standard oxygen atmospheric environment. This turned out to be true for all but one of the colonies.

One of the issues with the genetic models of Fisher and others is how to keep the benefits of an old gene needed for survival without replacing it with a new gene that could negatively impact survival. Fortunately, genomic studies showed that there are several mechanisms that meet the need

(Emlen and Zimmer, 2020, pp. 313–317). One of them is gene duplication. What happens is that an old gene is duplicated and added to the DNA. Then mutations to this new gene can add new capability without impacting the survival value of the old useful gene.

This is exactly what happened in Lenski's experiment with one of the colonies. By gene duplication and mutation, one colony could suddenly consume citrate (Blount et al., 2012). It could still consume glucose, but could now consume citrate also. The result was the dominance of this strain and the extinction of glucose-only-consuming strains. There was another glucose-consuming strain that survived, but only because it had mutated to also consume the detritus of the citrate-and-glucose-consuming strain (Turner et al., 2015). Nevertheless, the appearance of an evolutionary innovation by gene duplication that added capability and the subsequent extinction of unmodified competitors was a dramatic example of variation and selection.

Now consider the replicator model shown in Figure 1.3. In this model, a new resource R3 appears as shown in Figure 1.3a. Replicator 3 also appears as a variation of replicator 2 that consumes R2 and R3. Replicator 3 consumes R2 first. After R2 is consumed, then unfed members of population 3 consume R3. Detritus production is implicit. Note that replicators 2 and 3 are in competition for resource R2.

At this point, one could make a sampling model of this competition. The result would be that if the total population N=N2+N3 was less than or equal to the available resource R2, then each member of both populations 2 and 3 would be able to feed. But if N was larger than R2, then each population would sample the resource until it was depleted. If there was no sampling bias between the populations, then on average each population would get a share of the resource in proportion to its fraction of the total population N.

The model adopted in Figure 1.3b adopts the common physics and statistics approximation of replacing the sampling with average values (e. g., see Sanders and Verhulst, 1985) and is related to an approximation method in physics and mathematics called mean-field theory. This greatly simplifies the computations by replacing the need for averaging simulations with analytic computations. In general, it is expected that the results of using this approach will be the same as averaging sampling simulations.

Selection Model 1.3 Details. Note that in the formulas of Figure 1.3b, an intermediate variable N3P is introduced to compute the amount consumed by replicator 3 of the shared resource

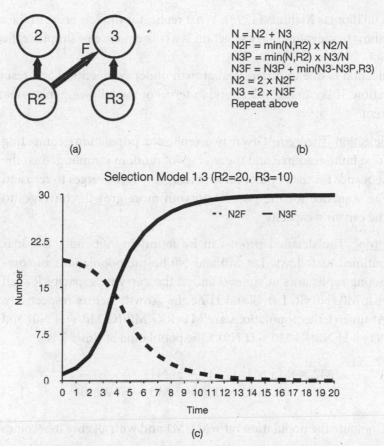

$$N = N2 + N3$$
$$N2F = \min(N,R2) \times N2/N$$
$$N3P = \min(N,R2) \times N3/N$$
$$N3F = N3P + \min(N3-N3P,R3)$$
$$N2 = 2 \times N2F$$
$$N3 = 2 \times N3F$$
Repeat above

(a)   (b)

Selection Model 1.3 (R2=20, R3=10)

(c)

FIGURE 1.3   Selection Model 1.3. (a) Food web. Replicator 3 is a variation of 2 that can feed on a new resource R3. However, 3 feeds on resource R2 first as indicated by the arrow with the letter "F." Then unfed 3 feed on resource R3. (b) Formulas. If the total number of replicators N=N2+N3 is larger than the shared resource R2, then the number fed (N2F and N3P) is approximated by an unbiased sampling average as shown. (c) Results. Replicator 2 is selected against because its share of resource R2 diminishes to nothing.

R2. Then N3P is totaled with unfed replicator 3 that consumes resource R3 to compute the total fed replicator 3.

Figure 1.3c shows the results of the model. Replicator 3 grows due to the new resource R3. Replicator 2 diminishes because its share of R2 decreases. Replicator 2 eventually goes extinct. This model illustrates the conventional view of how evolutionary selection works. The loser of a resource competition is selected against. This is classic Darwinian selection via the ideas of Thomas Malthus (1798). What replicator models bring to the table is to show that evolutionary selection is a consequence of simple replicator physics.

Selection is due to differential growth under competition and resource limitation. This can be understood in terms of the following mathematical theorem:

Selection Theorem: Given two replicator populations competing for a finite resource and the average of random sampling, then the sequence for the population with less growth converges to zero and the sequence for the population with more growth converges to the carrying capacity.

Proof: The detailed proof can be found in Appendix B but is outlined as follows. Let M0 and N0 be the populations of competing replicators at time=0 and at the carrying capacity R such that M0+N0=R. Let G and H be the growth factors respectively. At time=1 the populations are M1=R G M0/(G M0 + H N0) and N1=R H N0/(G M0 + H N0). One population at time=2 is

$$M2 = R \ G \ M1/(G \ M1 + H \ N1)$$
$$= R \ G^2 \ M0/ \ (G^2 \ M0 + H^2 \ N0) \qquad (1.5)$$

Compute the population ratio M2/M1 and with algebra it becomes

$$M2/M1 = 1 + (G - H) \ H \ N0/(G^2 \ M0 + H^2 \ N0) \qquad (1.6)$$

If G < H, then M2/M1 < 1. Since MK, the M population at time=K, can be written as a product of ratios

$$MK/M1 = (M2/M1) \ (M3/M2) \ . \ . \ . \ (MK/MK\text{-}1) \qquad (1.7)$$

and if G<H, the ratios are all less than one, as K increases MK goes to zero and consequently NK, the N population at time=K, goes to R. QED.

This is the classic essence of evolutionary selection. It doesn't matter what the causes are for growth. The Selection Theorem applies to all kinds of selection whether from extra resources, fertility, predation, pathogens, or sexual selection. The key factors are competition, resource limitation, and differential growth and the consequence is evolutionary selection. For example, if a predator consumes one prey species but not another competing prey species, this is reflected in the differential growth rate of the two prey populations. The same is true for differential growth effects on hosts by infectious diseases. Selection with predation and infectious diseases are discussed in greater detail in Chapters 5 and 7, respectively.

But selection takes time. Eqs. 1.6 and 1.7 reveal an important mathematical fact about the speed of selection:

Selection Speed Theorem: The speed of converging to zero or the carrying capacity is determined by the magnitude of the growth difference for replicator populations under competition for a limited resource and random-sample averaging.

Proof: The detailed proof can be found in Appendix B, but the key elements are Eqs. 1.6 and 1.7. As shown in Eq. 1.6, the population ratio M2/M1 decreases with decreasing growth rate G<H. The product of growth rates for MK/M1 in Eq. 1.7 also decreases and therefore approaches zero more quickly with the larger-growth population approaching the carrying capacity more quickly. QED.

The Selection Speed Theorem illuminates one of the central topics in evolutionary theory: the speed of evolution. Darwin envisioned evolution as a slow and gradual process but others have challenged that view (see e.g., Eldredge and Gould, 1972 and Hendry, 2017). The speed of evolution is clearly made up of two parts: the speed that variations appear and the speed of selection. If the variation required to take advantage of an opportunity like a new resource requires an evolutionary innovation, then the speed of variation appearing could be quite slow. For example, the citrate feeder in Lenski's *E. coli* experiment took over 31,000 generations to appear in one flask (Lenski, 2017) and it has yet to appear after 73,000 generations in any of the other 11 flasks (see Wikipedia, 2021).

But if the variation already existed relatively neutrally as in a sexually interbreeding population when an evolutionary opportunity appeared, then the speed of evolution is determined by the speed of selection. If the difference in growth rate is small, then selection could take many generations. But if, for example, an ecological change appeared that presented a new resource like that shown in Figure 1.3, the effective growth difference could be large and selection would proceed quickly. Thus, contrary to Darwin's view, the Selection Speed Theorem shows that evolution can proceed relatively both slowly and quickly.

It is worth noting that random-sample averaging is related to conventional ecology theory for competing species.

<u>Theory Detail.</u> Consider the logistic equations using Eq. 1.4 for two species populations NF1 and NF2 with different carrying capacities R1 and R2 and replication ratios K1 and K2, respectively. Set the carrying capacities equal to their random-sample averages for a common resource with carrying capacity R such that

$$R1 = R \, N1F/(N1F + N2F) \qquad (1.8)$$

$$R2 = R \, N2F/(N1F + N2F) \qquad (1.9)$$

Insert these into the respective logistic equations of Eq. 1.4 and with a little algebra one obtains

$$\Delta N1F = (K1 - 1) \, N1F \, (1 - (N1F + N2F)/R) \qquad (1.10)$$

$$\Delta N2F = (K2 - 1) \, N2F \, (1 - (N1F + N2F)/R) \qquad (1.11)$$

These are the difference equations of conventional ecology theory for species competition (see, e.g., Smith and Smith, 2006, p. 259). Consequently, the random-sample average leads to their derivation. The difference with replicator models is again the use of softer logistic density-dependent regulation for conventional ecology theory rather than the hard resource limits for the replicator models.

The Selection Theorem gives us an opportunity to discuss another important concept in evolutionary theory known as selection pressure. The basic idea is that any environmental factor that can cause evolutionary selection can be viewed as providing selection pressure. But the Selection Theorem provides a precise quantitative meaning to the concept of selection which is survival of competing populations from resource limitations and differential growth. Furthermore, it is important to determine how strong the selection pressure is because we expect that the strength of the selection pressure to determine the speed and probability of evolution.

A weak selection pressure would cause evolution to proceed very slowly or perhaps not at all, while a strong selection pressure would act quickly and dramatically. Note that the key factor in the proof of the Selection Theorem was the growth-rate difference. Therefore in order to estimate the magnitude of the effect of selection pressure, we define SPF.

Definition: The selection-pressure factor SPF is the normalized absolute value of the growth-rate difference between competing populations:

$$SPF = abs(G1 - G2)/NT \qquad (1.12)$$

where given two replicators with final populations for time=t of N1F(t) and N2F(t) we have

$$G1 = N1F(t) - N1F(t-1) \qquad (1.13)$$

$$G2 = N2F(t) - N2F(t-1) \qquad (1.14)$$

$$NT = N1F(t) + N2F(t) \qquad (1.15)$$

where, as before, the time scale is one generation.

This definition of the selection-pressure factor SPF has the virtue of being bounded ($0 <= SPF <= 1$), equal to zero for no growth-rate difference, and near 1 for total extinction in one generation for one replicator population but low population for the other. This is consistent with the expectation that no selection means no selection pressure and extinction means relatively larger selection pressure. Please note that if extinction occurs slowly that this is less selection pressure than if extinction occurs quickly and that this is reflected by the corresponding values of SPF. We shall show in the later chapters that selection pressure plays a very important role in the evolution of complexity and novelty.

Finally, let us consider again evolution at the beginning of life. Dawkins' idea was that evolution started with replication. So is there any evidence of this? Consider the metabolic pathways essential for cellular construction and gathering energy. Which came first, metabolism or replication? By testing many alternative possible pathways, it has been shown that the metabolic pathways of life produce the highest flux of compounds relative to alternatives (Court et al., 2015). This appears to show evolutionary selection at work. Because selection would have to predate metabolism, then replication which induces selection would have to occur first. This is consistent with the observation that protocells made of lipids with no metabolism can replicate (Frankel et al., 2014). It is the appearance of replicators that kickstarts evolution consistent with Dawkins' hypothesis.

Did primitive replicators occur before or after genetic machinery like RNA, DNA, and the alphabet genetic code of 20 amino acids? One study by Phillip and Freeland (2011) looked at a sample of 50 amino acids naturally found in meteorites and asked what set would have the important characteristics for building proteins such as a range of sizes, different charges, and a range of tendencies to repel water. Computer analysis of potential sets that have the best coverage across these characteristics showed that the basic set of 20 amino acids of terrestrial biology is the best as a toolbox for building proteins. This basic set appears to be selected by evolution. Thus, selection would seem to predate the alphabet of 20 amino acids. Consequently, the replication that causes selection would appear to occur before selection of the genetic code of the amino-acid alphabet.

1.4 COEXISTENCE

About 1.5 billion years ago, a tremendous evolutionary event occurred on Earth. At this time, the only living organisms were microbes from the domains of bacteria and archaea. Although these microbes can look similar, they have somewhat different genomes due to their ability to survive in different environments.

The great evolutionary event was that a bacterium got inside an achaeon and the bacterium evolved to become mitochondria (Lane, 2015). How it got inside is not known, but a clue is that one of the closest genomic relatives to mitochondria is an intracellular parasite (Emelyanov, 2001). At first, there must have been simple coexistence between the bacterium and the archaeon allowing both to survive. For example, intracellular parasites need to keep the host alive for as long as possible while growing and reproducing. But then some level of cooperation arose such that each gained some benefit. The archaeon provided nutrition while the early mitochondrion started to provide the energy molecule ATP. This enabled the archaeon to have much more energy and grow much more thereby providing it with a substantial evolutionary advantage. In the simple environment inside the archaeon, the mitochondria started to lose their genes and transfer them to the host. The host developed a nucleus and the evolutionary result was the full-scale symbiotic relationship known as the first eukaryote. The rest is history. The evolutionary power of this new eukaryotic organism created a new domain of life filled with multicellularity and complexity.

But before there was cooperation and symbiosis there had to be simple coexistence. Evolution by natural selection is based on competition, so how could coexistence appear? Consider what happens when the new replicator 3 in the previous replicator Selection Model 1.3 prefers to consume the new resource R3 first rather than last? What results is Figure 1.4a which is similar to Selection Model 1.3, but replicator 3 consumes resource R3 first.

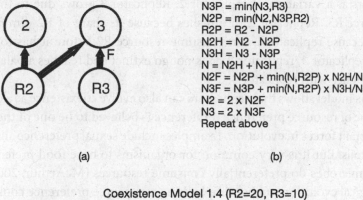

N3P = min(N3,R3)
N2P = min(N2,N3P,R2)
R2P = R2 - N2P
N2H = N2 - N2P
N3H = N3 - N3P
N = N2H + N3H
N2F = N2P + min(N,R2P) x N2H/N
N3F = N3P + min(N,R2P) x N3H/N
N2 = 2 x N2F
N3 = 2 x N3F
Repeat above

(a)   (b)

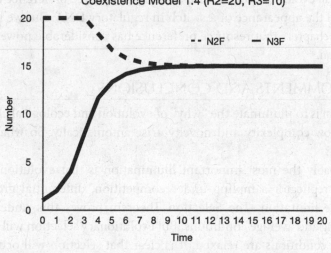

(c)

FIGURE 1.4   Coexistence Model 1.4. (a) Food web. This model is the same as Selection Model 1.3 except that replicator 3 consumes R3 first as shown by the arrow with the letter "F." (b) Formulas. Replicator 2 feed on R2 while replicator 3 feeds on R3. Then they share the remainder of R2. (c) Results. Replicator 2 does not go extinct but rather achieves a coexistence with replicator 3 as a consequence of resource preference.

Coexistence Model 1.4 Details. Replicator 2 feeds on resource R2 while replicator 3 feeds on R3. As shown in the formulas of Figure 1.4b, this first stage has a natural limit for N2 consumption that is set by whatever is the least of N2, R2, and N3P= min(N3,R3) which is the amount that replicator 3 consumes of R3. The remaining unfed replicators compete for the remainder of resource R2.

The results of this model are shown in Figure 1.4c. Again replicator 3 appears as a variation of replicator 2. Replicator 3 grows due to the new resource R3. Replicator 2 diminishes because its share of R2 diminishes. But because replicator 2 can consume resource R2 before it has to share with replicator 3, replicator 2 does not go extinct and reaches a balance of coexistence with replicator 3.

This model shows that replicators can also evolve coexistence as a consequence of resource preference. Preference is believed to be one of the most important forces in evolution. Examples include sexual preference and peacock tails. But it is very common for organisms to have food preferences. Even microbes do preferentially consume resources (McArthur, 2006). In biological evolution, such a simple thing as resource preference could be a result of the appearance of a switch in regulatory DNA. But we shall show in later chapters that resource preference has considerable power to shape evolution.

1.5 COMMENTS AND CONCLUSIONS

Our goal is to illuminate the "why" of evolution and ecological change and learn how complexity and novelty arise automatically. So what have we learned?

Probably the most important illumination is that evolutionary selection is replicator sampling under competition, differential growth, and resource limitation. The Selection Theorem proves this under random sampling and average conditions. But evolutionary selection will still occur if these conditions are relaxed. It is clear that selection will occur even if there are limited levels of biased sampling. If one population is getting an increasing share of the resource, then the competing population will be getting a decreasing share and will converge to zero. It is also clear that stochastic effects will play a role but that averages will win out given time. What makes evolutionary selection so powerful is its simple and universal character: just replication, competition, differential growth, and limited resources.

Evolution can proceed at different rates depending on ecological conditions. The Selection Speed Theorem proves this. If variations are slow in appearing, then evolution will be slow. But if the right variations are already present, ecological opportunities can appear that allow selection, and thereby evolution, to proceed relatively quickly.

The replicator models are related to conventional ecology theory. For example, the replicator models exhibit density-dependent regulation in a similar manner. In addition, conventional ecology theory for species competition can be derived from the formulas for random-sample averaging. But a major factor is the importance of resource preference and that needs a flexible approach like that of replicator modeling.

Metabolism and the genetic code are products of a selection process. Because replication induces selection, the appearance of replication in the origin of life would need to precede the appearance of metabolism and the genetic code. This supports Dawkins' hypothesis that evolution started with the appearance of replicators.

Detritus plays a critical role in evolution. Detritus is a natural consequence of resource limitation and it changes the ecology to provide new resources for the global expansion of evolution. The beautiful and awesome phenomenon of marine snow proves this. When variations first appear that can consume detritus, there will be an initial blooming effect that will eventually settle down to a conventional resource flow.

Coexistence comes easily to replicators. Coexistence is the critical forerunner to cooperation. And cooperation is central to the evolution of complexity. With the emergence of eukaryotes, biology opened the door to complexity by providing machinery sophisticated enough to build it. Coexistence and cooperation helps to walk through that door.

At the origin of life, the models are simple replicator models. But as evolution proceeds, more complexity is added and the models will reflect this. We will add features to our replicator models until they begin to resemble complex life. This will be the point where our simple replicator models will transition to more complex models. In addition, the technique of using random-sample averaging for shared resource consumption can be implemented as a simulation. However, by averaging the simulation results, we would get the models described which are very computationally amenable to solution.

CHAPTER 2

Fertility

IN THE BEGINNING, ALTHOUGH replicators were bound to the resource of origin, they competed with each other by using fertility to increase the number of their offspring. Fertility is one of the driving forces of evolution. In turn, variations in fertility and replication speed are also subject to evolutionary forces. In some circumstances, enhanced fertility is a very winning strategy. The Selection Theorem predicts this. Population growth over one's competition will lead to selection. In this chapter, we will examine several cases of the interaction of fertility and evolution. Such cases include juvenile mortality, invasive species, streamlining, multicellularity, and evolvability. But there are important limitations to this strategy. This chapter discusses both the advantages and disadvantages of enhanced fertility in evolution.

2.1 JUVENILE MORTALITY

When life began, juvenile mortality must have been considerable. Between copying errors and environmental hazards, the chances of survival must have been slim. Unless there are enough offspring to beat mortality, a species will decline and disappear. This is especially true because of juvenile mortality. It has been a major factor for evolution and enhanced fertility has been a major mechanism for compensating.

DOI: 10.1201/9781003391395-2

After the emergence of the first replicators, it is clear that the error rate of replication would have been a challenge. Juvenile mortality must have been fearsome and if the mortality rate exceeded the replication ratio, the replicators would have gone extinct. What was needed was some kind of memory that was passed on to the next generation. Initially, there was some structural memory, like the constraints on metabolism and protocells (Goldford et al., 2019). But any additional memory that lowered juvenile mortality would be selected because of the improved growth rate. So what must have occurred was that information machinery must have evolved step by step to make this adaptation. A likely scenario is first simple RNA followed by more complex RNA (Neveu et al., 2013) and then a combination of DNA and RNA. The improved information system would have been selected because of the improved growth rates.

Examples of evolutionary phenomena can often be found in current and recent history even among human populations. This has also been true with juvenile mortality. One of the great achievements of the twentieth century was the very large reduction in the mortality rate of children. Before the twentieth century, the mortality rate of children was an absolutely horrific 46.2% (Volk and Atkinson, 2013). The causes were disease, congenital problems, violence, and accidents. But after improvements in sanitation, water quality, food safety, and medical care, the developed world greatly reduced the rate to about 1%.

But history also showed that there was considerable variability in children's mortality. For example, the rate for Finland during 1749–1773 was 35% while for Germany during 1692–1899 it was 60%. The rate for modern hunter-gathering peoples varies from 22% to 58.5% although the average of 48.8% is similar to the historical general rate before the twentieth century. Nonhuman primates have similar rates with juvenile mortality varying from 27% for orangutans to 67% for lemurs.

There is no reason to expect that this variability in juvenile mortality would not exist throughout almost all of historical biology. In particular, at the early stages of life, juvenile mortality must have been large until early microbes evolved sufficiently to have machinery that would allow them to be reasonably adapted to their environmental niches. But just like modern biology, environmental variations would still be expected to create variations in juvenile mortality.

But the variability of juvenile mortality raises the evolutionary question of what would happen if two populations with different rates competed

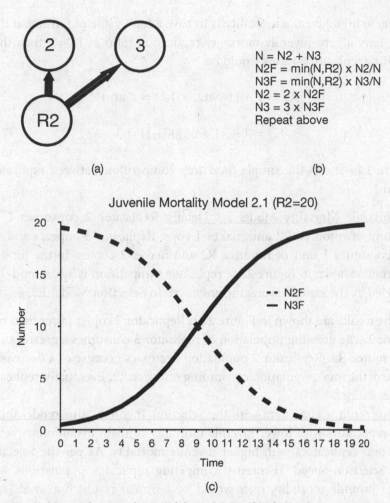

N = N2 + N3
N2F = min(N,R2) x N2/N
N3F = min(N,R2) x N3/N
N2 = 2 x N2F
N3 = 3 x N3F
Repeat above

(a)   (b)

Juvenile Mortality Model 2.1 (R2=20)

-- N2F
— N3F

Number

Time

(c)

FIGURE 2.1   Juvenile Mortality Model 2.1. (a) Food web. Replicator 3 appears and is similar to replicator 2 but with lower juvenile mortality. Replicator 2 and replicator 3 share resource R2. (b) Formulas. Replicators 2 and 3 have juvenile-mortality rates of 2/3 and 1/3 with corresponding replication ratios of 2 and 3, respectively. The sampling-average approximation is again employed for consumption competition. (c) Results. Replicator 3 population rises because of the increasing share of the resource. Replicator 2 population falls to eventual extinction because of its higher juvenile mortality and the consequent decreasing share of the resource.

for the same resource. Consider the model of two competing replicators shown in Figure 2.1.

Replicator 2 has a juvenile-mortality rate of MJ2=2/3. Replicator 3 has a juvenile-mortality rate of MJ3=1/3. One of the ways that evolution adapts to high juvenile mortality is to have a high value of fertility. If they both have a zero-juvenile-mortality replication ratio of K0=4, then their resulting replication ratios would be

$$K2 = 1 + (1-MJ2)(K0-1) = 2 \text{ and} \tag{2.1}$$

$$K3 = 1 + (1-MJ3)(K0-1) = 3 \tag{2.2}$$

Figure 2.1a shows the simple food-web competition between replicators 2 and 3.

> Juvenile Mortality Model 2.1 Details. Replicator 2 consumes 1 unit of resource R2 and makes 1 copy. Replicator 3 appears and consumes 1 unit of resource R2 and makes 2 copies. In the formulas shown in Figure 2.1b, replicator competition is again modeled by the sample averaging employed in Selection Model 1.3.

The results are shown in Figure 2.1c. Replicator 3 copies faster than replicator 2. The growing population of replicator 3 consumes a greater share of resource R2. Replicator 2 population decreases because of a decreasing share of the total population consuming resource R2. Eventually, replicator 2 goes extinct.

This result is consistent with the Selection Theorem. This model shows that replicators with lower juvenile mortality can win evolutionary selection over replicators with higher juvenile mortality. As per the Selection and Selection-Speed Theorems, competing replicator populations with closer juvenile-mortality rates would have similar results but would take more generations to reach selection and extinction. Nevertheless, it is expected that evolutionary selection of reduced juvenile mortality would play a strong role in the selection of early life forms.

2.2 INVASIVE SPECIES

One of the most damaging modern pests is invasive species. For example, Hawaiian forests have been destroyed by feral pigs. Freshwater clams in lakes and rivers have disappeared due to Zebra mussels. Cattle grasslands have been taken over by inedible leafy spurge. The list goes on and on of invasive species that have wrecked ecological and economic havoc (Van Driesche and Van Driesche, 2000).

But invasive species have also played a strong role in evolutionary history. The most dramatic example is the Great American Interchange that took place starting about 3 million years ago (Ridley, 2004, pp. 512–517). North and South America were separate continents before this time. Both had developed distinctive faunas that were suited to their different environments. The south had heavily armored armadillos, marsupial saber-toothed carnivores, giant ground sloths, and giant rats. North America had dogs, cats, and bears.

Then the modern Isthmus of Panama was formed that connected the two continents. Dogs, cats, and bears invaded from the north, and armadillos, opossums, and marmosets invaded from the south. What resulted was mostly a takeover of the south by the north that was aided by the weight of numbers and relative success. One important factor was that the northern species appeared to have evolved a more demanding competitive existence. For example, they had more advanced armaments that allowed them to overrun the southern mammals. The result was that about 50% of southern mammals now have a northern origin. And the marsupial saber-toothed carnivores, giant ground sloths, and giant rats have all gone extinct.

One of the ways that invasive species can take over is to simply replicate faster than the native species. However, more consumption of resources is required to do that. Model 2.2 is a modification that reflects the added consumption needed for greater fertility. In particular, replicators need to consume one resource unit per copy.

> Invasive Species Model 2.2 Details. As shown in Figure 2.2a, replicator 2 consumes 1 unit of resource R2 and makes 1 copy. Replicator 3 appears and is an invasive specie. Replicator 3 consumes 2 units of resource R2 and makes 2 copies. A slight reformulation is needed to accommodate the extra consumption. The potential resource consumed is N=N2+2xN3. If N<R2, then there is plenty to go around and N2F=N2 and N3F=N3. But otherwise R2=N2F+ 2xN3F. The sampling-average approximation maintains sampling share so N2F/N3F = N2/N3. With a little algebra we get for N>=R2, N2F=R2xN2/N and N3F=R2xN3/N and the final formulas shown in Figure 2.2b. Thus the replicator competition for resource R2 is modeled by the sampling-average approximation reformulated for 2-unit consumption.

The results are shown in Figure 2.2c. Replicator 3 copies faster than replicator 2. The growing population of replicator 3 consumes a greater share of resource R2. Replicator 2 population decreases because of a decreasing

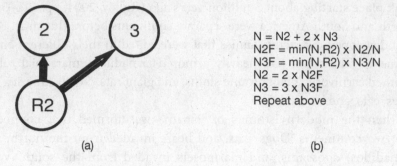

N = N2 + 2 x N3
N2F = min(N,R2) x N2/N
N3F = min(N,R2) x N3/N
N2 = 2 x N2F
N3 = 3 x N3F
Repeat above

(a)   (b)

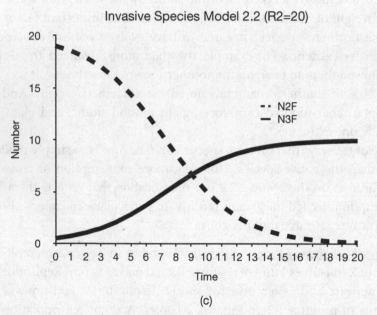

(c)

FIGURE 2.2   Invasive Species Model 2.2. (a) The food web shows replicator 3 appearing as an invasive species against replicator 2 with consumption of one resource unit per replicator-unit copy. (b) Formulas are modified for 2-unit consumption of replicator 3 with the derivation in the text. (c) Results show that higher fertility invasive species win the competition for resources against lower fertility.

share of total population consuming resource R2. Eventually, replicator 2 goes extinct. Replicator 3 population goes to its final value due to consumption of 2 vs 1 resource units per time cycle. The model again shows that invasive species with higher fertility can win the competition for resources against lower fertility.

One may wonder if the Selection and Selection-Speed Theorems apply if there is differential consumption of resources for competing populations as in Model 2.2. The answer is yes, and the following theorem applies.

Extended Selection Theorem: The Selection and Selection-Speed Theorems are valid for differential consumption.

Proof: The detailed proof can be found in Appendix B but is outlined as follows. Let M0 and N0 be the populations of competing replicators at time=0. Let U and V be the resource consumption per replicator per unit time of the respective populations. Let the carrying capacity R be such that U M0+ V N0=R. Let G and H be the growth factors respectively. At time=1 the populations are M1=R G M0/(G U M0 + H V N0) and N1=R H N0/(G U M0 + H V N0). Compute the population M2 at time=2 and ratio M2/M1 and with algebra it becomes

$$M2/M1 = 1 + (G - H) H V N0/(G^2 U M0 + H^2 V N0) \quad (2.3)$$

If G < H, then M2/M1 < 1. The remainder of the proofs follows as before. QED.

Even with differential consumption, selection is a result of competition, resource limitation, and differential growth.

2.3 STREAMLINING

During the great pioneer migrations of the American West, the wagons were filled with things the pioneers thought they would need in their new homes. China sets, spinning wheels, clothes trunks, and memorabilia were hauled along. But the trip was long and arduous. The horse and oxen teams began to falter pulling their wagons over rough and arid terrain. Eventually, the pioneers had to lighten the wagons and discard their contents. The trail was lined with abandoned possessions and the graves of those who didn't make it. When they finally arrived, often all that was left was the wagon, a much smaller horse or oxen team, themselves, and very little else. The pioneers had to streamline to survive.

As with human history, so too it is with evolution. There are very special bacteria that live in the sea on very little nutrients known as *Pelagibacter ubique* or *P. ubique* for short. The name literally means bacterium of the sea that is ubiquitous. *P. ubique* is probably the most numerous bacteria in the ocean and even on Earth. It feeds on detritus and undergoes seasonal cycles reaching about 50% of the cells in temperate ocean waters. Thereby, it plays a major role in Earth's carbon cycle.

But what makes *P. ubique* truly exceptional is that it has the smallest genome of any free-living organism. Its genome has been streamlined by reducing the amount of energy required for cell replication (Giovannoni et al., 2005). The only cellular species with smaller genomes are intracellular symbiotes or parasites such as mitochondria or their closest genomic relatives. However, there now are several examples of genome streamlining which include both mitochondria and cyanobacteria.

So what could account for this evolutionary streamlining? Consider what would happen if the fertility of a new variation is only slightly larger than the ancestor population? Figure 2.3 examines the consequences of that. The food web shown in Figure 2.3a is the same as for Models 2.1 and 2.2.

> Streamlining Model 2.3 Details. Replicator 2 consumes 1 unit of resource and makes 1 copy. Replicator 3 appears and is a new variation of replicator 2. Replicator 3 also consumes one resource unit per unit copy. Replicator 3 replicates only slightly faster than replicator 2 and so consumes 1.1 units of resource R2 and makes 1.1 copies on average per unit time. Replicator competition for resource R2 is modeled by the sampling-average approximation reformulated for 1.1-unit consumption. The reformulation shown in Figure 2.3b follows the same derivation method described for Model 2.2.

The results are shown in Figure 2.3c. Replicator 3 replicates slightly faster than replicator 2 (replication ratios of 2.1X vs 2X). The growing population of replicator 3 consumes a greater share of resource R2. Replicator 2 population decreases because of the decreasing share of total population consuming resource R2. Replicator 2 takes longer to eventually go extinct. Replicator 3 population goes to a smaller limit compared to Model 2.1 due to consumption of 1.1 vs 1 resource units per time cycle on average. This model shows that even incremental increases in fertility can win the competition for resources. This is consistent with the Extended Selection Theorem which shows that even small differences in growth will lead to selection.

But how does streamlining lead to differential fertility and growth? Consider the following simple model of fertility potential. Let RS be the resources required to sustain a replicator per unit time without replication. Let RR be the resources required for a single replication. And let RC be the resources consumed per replicator per unit time. If the resources available for replication (RC-RS) equals the amount expended for replication (RR(K-1)) where again K is the replication ratio, then

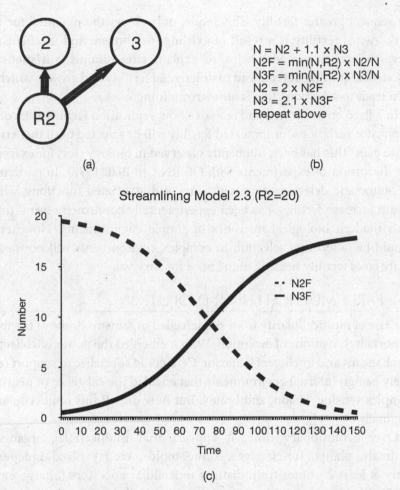

$$N = N2 + 1.1 \times N3$$
$$N2F = \min(N,R2) \times N2/N$$
$$N3F = \min(N,R2) \times N3/N$$
$$N2 = 2 \times N2F$$
$$N3 = 2.1 \times N3F$$
Repeat above

(a)   (b)

Streamlining Model 2.3 (R2=20)

(c)

FIGURE 2.3   Streamlining Model 2.3. (a) The food web is the same as Models 2.1 and 2.2 but with only incrementally larger replication ratio for replicator 3 (2.1 vs 2) and again the rule of one unit of resource consumption per copy. (b) The formulas are similar to Model 2.2 but with new incremental replication ratio. (c) Results show that even with an incrementally faster fertility, replicator 3 wins the competition although it takes more time.

$$K = 1 + (RC - RS)/RR \qquad (2.4)$$

In simple environments, this model explains how replicators with larger replication ratios can arise. Replicators with variations that decrease resources RS needed for sustainment and/or resources RR needed for replication can have larger replication ratios K and consequent greater fertility. Greater resource value (i.e., nutritional) for RC can also have larger K and consequent greater fertility. This model only shows the potential for fertility. Actual fertility is a result of existing metabolism and mechanisms. Nevertheless, this model helps to explain streamlining. Variations in resource requirements can lead to differential fertility and growth which in turn leads to selection. This is how streamlining works.

In a fixed environment and resource flow, replication fertility is favored. Replicator variations for increased fertility will be selected until there is no more gain. This has been frequently observed in biology (see, for example, the discussion of experiments with Qβ RNA in Bell, 1997). In particular, the automatic deletion of unused genes and associated functions, which would increase fertility, has been experimentally confirmed several times with modern biological microbes in simple environments. However, it should be noted that selection in complex environments will not necessarily favor fertility because of the need for survival.

2.4 EARLY MULTICELLULAR EVOLUTION

The rise of multicellularity from single-celled organisms is one of the most important transitions of evolution. With it emerged the power of collective mechanisms and intelligent behavior. Cells could specialize to support relatively benign internal environments that enabled the existence of neurons, complex sensing, vision, and brains. But how did all this multicellularity originally evolve?

Over evolutionary time, it appears that multicellular organisms (animals, plants, fungi, algae, slime molds, etc.) evolved independently at least 25 times from distinct unicellular ancestors (Sharpe et al., 2015). The largest and most familiar multicellular organisms evolved through cells dividing but not separating. This was followed by the differentiation of cell lineages into different specialized types. This kind of "multicellularity-by-division" evolved in simple organisms with a few cell types such as the green algae *Volvox* (Herron and Nedelcu, 2015). It also evolved in complex organisms such as animals with hundreds of cell types (Rokas, 2008).

The *Volvox* group comprises both unicellular species and multicellular species with various numbers and types of cells. The most common habitat is warm, nutrient-rich, and predator-sparse freshwater ponds and pools. The origin of multicellularity in this group was probably much more recent than those of complex multicellular animals, land plants, and fungi. For our purposes, the relevant feature of this group is the existence of extant species with nearly every conceivable degree of complexity from single cells to differentiated multicellular organisms. *Volvox* is one of the many ancient multicellular inventions. However, the history of volvocine evolution includes several well-supported instances of multiple origins and reversals (Herron and Michod, 2008). In addition, its evolution did not give rise to higher plants with anything more complex than *Volvox* itself (Bonner, 2016).

Nevertheless, the *Volvox* group provides an excellent model system for experimentally investigating the size-related advantages that might have caused single-celled volvocine algae to evolve multicellularity. In one experimental program, varying nutrient concentrations and predation rates from a unicellular predator produced variable growth rates for unicellular and multicellular *Volvox*. As a result, the analyses supported the hypothesis that predation was an important selective pressure for the origin of multicellularity in *Volvox* (Solari et al., 2015).

So what can the physics models tell us about the early evolution of multicellular organisms from single-celled ones? Consider the model shown in Figure 2.4a. This has the same food web as Invasive Species Model 2.2, but the new formulas in Figure 2.4b are a generalization of arbitrary resource consumption V for replicator 3 and arbitrary replication ratios K2 and K3. The parameter V accounts for consumption of multiple units of resources for a multi-celled replicator.

Now consider what happens when a two-celled replicator 3 appears and competes with the population of one-celled replicators 2. We can set V=2 for the consumption needs for each of the two cells for replicator 3. Because adhesion of the two cells via say an extra-cellular matrix also takes up some resources, we also consider the case of V=2.5 as an example. The results of equal replication ratios K2=K3=2 are shown in Figure 2.4c and there is no change in population numbers. This is consistent with the Extended Selection Theorem. If the two-celled replicator 3 has exactly the same growth rate or replication ratio as the one-celled replicator 2, there is no selection. The variation for adherence of two cells into one organism is a neutral variation if the growth rate is the same.

However, in the physical world, it is unlikely that the unicellular and multicellular growth rates are identical. Unicellular organisms take up needed resources through one or more pores in the cell wall. When two cells adhere together, if there is any blockage of those pores, then the resource uptake will be reduced. With fewer resources per time, the replication

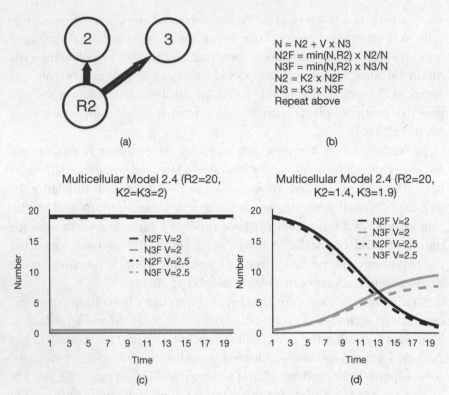

(a)

N = N2 + V x N3
N2F = min(N,R2) x N2/N
N3F = min(N,R2) x N3/N
N2 = K2 x N2F
N3 = K3 x N3F
Repeat above

(b)

(c)

(d)

FIGURE 2.4 Multicellular Model 2.4. (a) The food web shows replicator 3 appearing as a multicellular species against replicator 2 with consumption of one resource unit per replicator 2 and V units per replicator 3 copy. (b) Formulas are modified for V-unit consumption of replicator 3 with the derivation following Invasive Species Model 2.2. (c) Results show that multicellularity is neutral for equal growth rates. (d) Multicellularity is selected due to an increased growth rate as a consequence of more successful survival in a stressful environment.

rate and the corresponding growth will be reduced, and the multicellular organisms will be selected against. Not just neutral selection, but negative selection will occur against multicellular variations. Clearly, something else must be going on to select for multicellularity.

Here is where ecology and ecological change can provide selection pressure for multicellularity. Consider if replicator 2 is under some kind of environmental assault so that its growth rate and replication ratio are suppressed. This reduction in growth could come from significant mortality from numerous factors including parasites, pathogens, and predation. Now generalize the Figure 2.2b formulas from Invasive Species Model

2.2 to include arbitrary mortality M2 and M3 for replicators 2 and 3 with zero-mortality replication ratios of K20 and K30.

$$K2 = (1-M2) \, K20 \text{ and} \qquad (2.5)$$

$$K3 = (1-M3) \, K30 \qquad (2.6)$$

If the mortality M3 for the multicellular-variation replicator is less than the mortality M2 for the unicellular replicator and the zero-mortality replication ratios are the same, then the Selection Theorem predicts that the multicellular variation will be selected. The increased size could help reduce mortality by reducing surface area per volume, which is advantageous against parasites and pathogens. Another advantage that reduces mortality is being too large for predator ingestion. Thus, the onset of multicellularity is analogous to defensive herding in prey animals (Hamilton, 1971).

Figure 2.4d illustrates the effect of reduced mortality in showing the results for K2=1.4 and K3=1.9. For K20=K30=2, this corresponds to mortalities of M2=30% and M2=5%. Multicellularity emerges because it survives better than single cells in a challenging environment. Variations with more cells can appear over time because the environment of parasites, pathogens, and predators can also change over time as well. This topic is discussed further in Chapters 5 and 7.

However, if parasites, pathogens, and predators should disappear from the environment, there would be selection pressure to streamline as described in the previous section which could offer a hypothesis for the evolutionary reversals in the volvocine lineage. This is not an improbable scenario because if, for example, a predator doesn't adapt fast enough, it could go extinct, the environment would revert to a much less challenging one, and unicellular organisms could have the advantage. The relative isolation of freshwater ponds and pools would be an enabling factor.

2.5 EVOLVABILITY

The great and sometimes tangled tree of life tells the story of the rise of species over biological history. But it also tells another story—the appearance of the great innovations that made the large tree branches possible. The great innovations include DNA, photosynthesis, complex cells, sex, movement, sight, warm blood, and consciousness (Lane, 2009). But the greatest innovation of all was evolvability: the power to change the direction of evolution.

Evolvability is the ability to adapt. For example, bacteria have high evolvability. Their development of antibiotic resistance and the ability to eat plastic is a testament to this. Yet for all their adaptability, bacteria are limited by energy constraints that confine them to a single-cell existence. It was the great endosymbiosis event with the appearance of mitochondria and eukaryotes that took a quantum leap in energy availability and subsequent evolvability. The rise of the multi-cell tree of life is a testament to this.

Conversely, certain proteins seem to have come to the end of the line with respect to evolvability. These proteins have appeared to evolve optimal configurations and attempts to alter them just make them worse (Tokuriki and Tawfik, 2009).

Evolution has benefited from numerous key innovations that not only enabled survival but also passed on the ability to adapt better for future generations. This is the essence of evolvability—not just survival but survival enhancement for entire lineages. Evolvability is at the heart and soul of evolution.

So what does fertility have to do with evolvability? Consider Model 1.1 at carrying capacity, but this time with a replication ratio of K. Then the carrying capacity is N1F=R1 and the total population after replication is N1=KxR1. At the next cycle of feeding, only R1 of N1=KxR1 are fed and the rest do not survive. That means that only 1/K of the population survive. That also means that the mortality rate M is

$$M=1-1/K=1-1/(1+B) \tag{2.7}$$

which increases with increasing number of offspring B=K-1. Higher fertility means higher mortality at carrying capacity. A classic observation of Thomas Malthus.

But this relationship between mortality and fertility also suppresses evolution. Unless a new variation is linked to a new resource, that variation has the same chance for survival as the rest of the population, that is 1/K. The higher the fertility, the smaller the odds are for a variation of this kind to be successful. Many biological species have high fertility rates. But they pay a price of a lower ability to evolve.

Is there biological evidence of this relationship between fertility and evolvability? If we could count the number of variations per species and compare that to B, the number of offspring, that might be indicative. Over time, we might expect that

$$\#Variations \sim MutationRate \times Time \times SurvivalRate \qquad (2.8)$$

So if time and mutation rate are close for a group of species, then the number of variations could be proportional to $1/(1+B)$.

Evolutionary development (evo-devo) tells us that the main action in variations is not the protein-coding DNA segments, which is about the same in many species, but rather the regulatory DNA or switches. If we could count the number of DNA switches per species and compare that to B, that would be very interesting. The number of switches has not been counted yet, but if a species' physical size is driven by the number of switches, a not unreasonable hypothesis, then the physical size could be approximately proportional to the number of variations. Given the #Variations formula above, is there evidence that physical size is proportional to $1/(1+B)$?

Consider Figure 2.5 which uses data for mammals taken from Charnov and Earnest (2006).

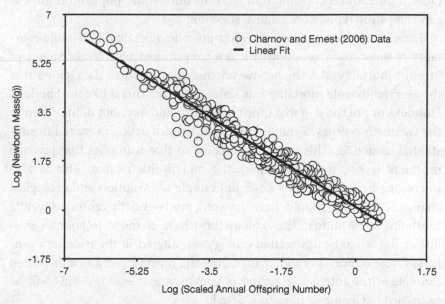

FIGURE 2.5   Offspring size vs offspring number in mammals based on data from Charnov and Ernest (2006). Per their analysis, the offspring number is scaled by female body mass raised to the 0.75 power to account for reproductive allocation per unit time scaling across species. The linear fit slope is about −0.92 which is not inconsistent with the evolvability model described in the text.

They analyzed the relationship between offspring number and size in newborn mass for mammals. There is a nearly inverse relationship between size and the number of offspring B:

$$Size \sim 1/B \qquad (2.9)$$

This is very close to what has been predicted. Obviously, additional analysis is needed, but the biological evidence is not inconsistent with the prediction.

2.6 COMMENTS AND CONCLUSIONS

In this chapter, we have discussed various aspects of fertility in relation to evolution, ecology, and complexity. So what have we learned?

First of all, the Selection Theorems can be extended to arbitrary differential consumption. Differential consumption just means that one replicator eats more than another replicator. But this also means that unless there is a change in population growth, selection is neutral with respect to consumption. Just eating more only counts only if your population grows more. Only if differential consumption leads to differential population growth can competing replicators achieve selection.

There is selection pressure to lower juvenile mortality. If juvenile mortality is static, then selection will favor replicators with lower values of juvenile mortality. But the pre-twentieth-century human data shows that the average juvenile mortality was about constant. This is like the Sherlock Holmes story of the dog that didn't bark. It should have but didn't. Then in the twentieth century, a number of health-related actions resulted in substantial reduction. This is very interesting in that it implies that juvenile mortality is not static but interactive and health related. This is also interesting because we now have an example of evolution and ecological change. Evolution should have lowered pre-twentieth-century juvenile mortality, but it didn't. The ecology must have changed to counter evolution. But when health-related ecology was altered in the twentieth century, juvenile mortality did change. Juvenile mortality is a health-related example of the interactive nature of evolution and ecology. This will be examined in Chapter 7 in greater detail.

Invasive species can displace native species by having greater fertility. So why aren't invasive species a greater scourge than they have been? Clearly, geographical barriers have played a part in keeping ecologies separated.

But even before modern travel exacerbated invasive species, geographical barriers were porous. For example, even before the opening of the Isthmus of Panama and the Great American Interchange, substantial species invasion was accomplished by island hopping.

So that still begs the question of why there isn't more species invasions. The answer is suggested by the case of leafy spurge and cattle grasslands. When leafy spurge arrived in the Great Plains from Eastern Europe, it initially had no natural enemies to control it. But when ranchers discovered this, they introduced the natural enemy, the flea beetle, and the leafy spurge abated. This again is an example of the interactive nature of evolution and ecology. Invasive species are definitely an evolutionary force, but changing the ecology by the arrival of natural enemies changes the evolutionary dynamics. This interactive phenomenon is explored in greater detail in Chapter 5.

Simple environments can evolve simpler replicators by streamlining. This is accomplished because variations that enhance fertility by more effective use of resources and subsequent possible genome loss will cause populations to outgrow their competitors as long as the environmental conditions don't change. But what happens if the environment does change? The next chapter explores this.

For early multicellularity, a variation containing simple adhesion of two cells from single-celled ancestors is not enough to evolve two-celled organisms without some improvement in growth rates. But a challenging environment for which multicellularity has an advantage will work. However, multicellular organisms can slide back to unicellular ones if the challenge goes away.

Finally, we have discussed the effects of fertility on evolvability. The problem is that the carrying capacity creates mortality and more offspring means more mortality. This means less survivability for a new variation unless it has a significant selection advantage like consumption of a new resource. Thus fertility can suppress evolvability. There is some suggestive evidence for this in size versus clutch size in mammals. However, the best evidence to support this will come when regulatory genes can be counted, totaled with protein-coding genes, and compared across the fertility of species.

CHAPTER 3

Scavengers

AFTER LIFE BEGAN ON Earth, the ocean floor began to be covered in marine snow. At first, it was just small amounts near the hydrothermal vents where replicators thrived. This marine snow was made up of organic detritus and so did not melt but rather built up layer upon layer, year after year, millennia after millennia, for many millions of years. These layers contained the nutrients for potential life.

And so eventually replicator variations appeared that could scavenge on the marine snow. But the nutrient value of marine snow varied with its age. Younger marine snow still had many of the organics of life while older snow had become more basic organic compounds. These differences in nutrient value also enabled the appearance of other replicator variations that could take advantage of their different nutrient value. Some replicators were able to find regions where the environment didn't change much. These replicators specialized in consuming marine snow of a certain age. Thus specialists arrived to take advantage of these relatively simple environments.

But where multiple ages of marine snow were available, replicator variations appeared that could scavenge more broadly and take advantage of the resource variations. And so the first generalists appeared. Rather than specialize in one kind of resource, they adapted to consume several kinds and became generalists. Modern examples of specialists are koalas while generalists include raccoons. Eventually, the specialists and the generalists interacted, and competition, selection, and coexistence resulted. This interaction of specialists and generalists is at the foundation of complex ecologies. In this chapter, we shall explore the mechanisms of this interaction using the tools of replicator models in order to illuminate the early scavenger phase.

3.1 SPECIALISTS

In 1934, Georgyi Gause published the results of experiments of competing unicellular organisms (Gause, 1934). The experiments with Paramecium involved culturing together two species, *P. aurelia* and *P. caudatum*, in a flow of water and nutrients. Although they initially grew and coexisted for a while, eventually *P. aurelia* drove *P. caudatum* to extinction. Gause contended that two species with similar ecology could not live together in the same place. This concept became known as the "competitive exclusion principle" or Gause's Law (Hardin, 1960). Simply stated, it says that complete competitors cannot coexist.

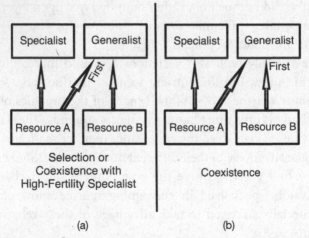

(a)

(b)

FIGURE 3.1 Specialist-versus-generalist-tradeoff model. Specialists consume resource A while generalists consume resources A and B. (a) The generalists consume the shared resource A first. Selection Model 1.3 shows that this can result in selection of the generalists and extinction for this specialists. See Invasive Specialist Model 3.2 for the potential for coexistence. (b) Because the generalists prefer the resource B that is not shared, the result is coexistence with the specialists as shown in Coexistence Model 1.4.

It is interesting to consider the exclusion principle in the context of competing replicator specialists. The fertility models of the previous chapter show that if there is any difference in growth rates, the faster specialist will win and the slower specialist will go extinct. This is completely expected by the Selection Theorem. So in the case of specialist-versus-specialist competition, competitive exclusion is merely evolutionary selection.

But when specialists compete against generalists, a more complex picture emerges. Selection Model 1.3 and Coexistence Model 1.4 can be considered from the tradeoff perspective shown in Figure 3.1.

The specialists consume only one resource while the generalists more than one. If the generalist consumes the shared resource first, Selection Model 1.3 showed that the specialist can go extinct. Whereas if the generalist consumes the shared resource last, then Coexistence Model 1.4 showed that coexistence can result. The addition of the unshared resource is outside of the conditions necessary for competitive exclusion. However, this shows the importance of resource preference in specialist-generalist interactions. The next section shows that under certain conditions, Selection Model 1.3 can result in coexistence.

3.2 INVASIVE SPECIALISTS

When silver carp, a variety of Asian carp, was introduced to North America in the 1970s, the intent was to use them to control algae in fish ponds. They are voracious feeders of phytoplankton and can reach sizes of 55 inches and 110 pounds. Because they are also very prolific, they have been useful in controlling algae blooms. One peculiar feature is that when they are startled, they tend to leap out of the water up to 10 feet in the air. Thus, water skiing around silver carp is extremely dangerous.

Unfortunately, whether by accident or intent, the silver carp entered the waterways of North America. Because of their prolific and voracious behavior, they have had a negative impact on native fishes and so have been declared invasive species. There is considerable concern that silver carp will enter the Great Lakes and negatively impact commercial fishing. They have already appeared to have had a negative impact on the native Bigmouth Buffalo fish (Phelps et al., 2017). One interesting feature of the interaction between silver carp and Bigmouth Buffalo is that adult silver carp feed almost exclusively on phytoplankton and so can be considered as a specialist. However, Bigmouth Buffalo feed not only on phytoplankton but also on small crustaceans and insects and so can be considered as a generalist.

Consequently, let us consider the case of an invasive specialist versus a generalist to learn what replicator models can tell us. In Chapter 1, the Selection Model 1.3 specialist replicator 2 was selected against because generalist replicator 3 won the competition of sharing resource R2 by consuming the shared resource first and being able to increase its population with consumption of the new resource R3. However, we have shown in Chapter 2 that competitions can also be won by increasing fertility. What would be the consequences of an interaction between increased fertility and resource competition?

We shall continue to postulate only small variations for our thought experiments. The appearance of a replicator with a large difference in fertility would not be a small variation. However, there are also invasive species. This is where small variation steps occur in a different location and the result invades the ecology. Note that although change can be large locally from invasive species, they too would only evolve from an accumulation of small steps.

Consider Figure 3.2a with a food web that is almost the same as Selection Model 1.3. But this time, specialist replicator 2 is an invasive species, indicated by the "+" sign, that specializes in consuming R2 by having twice the replication ratio.

> Invasive Specialist Model 3.2 Details. Generalist replicator 3 consumes R2 first, as indicated by the arrow with the letter "F." As shown in the formulas in Figure 3.2b, consumption of resource R2 by replicators 2 and 3 employs the usual sampling-average approximation. Again, generalist replicator 3 then feeds on resource R3.

The results are shown in Figure 3.2c. Specialist replicator 2 does not go extinct as before and the population grows due to increased fertility. It is the specialization in consuming R2 by having larger fertility that gives replicator 2 the advantage. In this case, replicators 2 and 3 reach a balance and coexist. This model demonstrates once again that fertility can play a strong role in evolutionary selection.

We have shown that at some specialist replication ratios, the specialists can go extinct when they interact with generalists (Model 1.3). At other replication ratios, they can coexist (Model 3.2). What is the basis for this phenomenon and what does it mean relative to the Selection Theorem?

Let K2 and K3 be the replication ratios for specialist replicator and generalist replicator 3 respectively. Remember that R2 and R3 are the resource

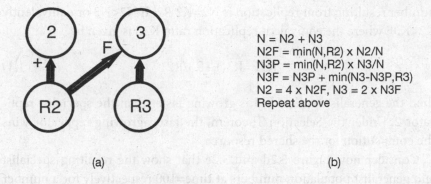

$N = N2 + N3$
$N2F = min(N,R2) \times N2/N$
$N3P = min(N,R2) \times N3/N$
$N3F = N3P + min(N3-N3P,R3)$
$N2 = 4 \times N2F, N3 = 2 \times N3F$
Repeat above

(a)   (b)

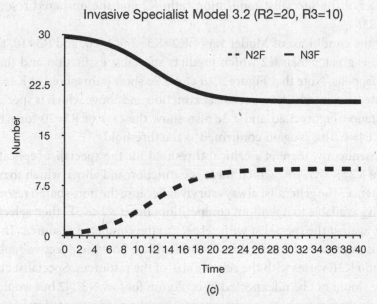

(c)

FIGURE 3.2   Invasive Specialist Model 3.2. (a) The food web shows an invasive specialist replicator 2, indicated by the "+" sign, that competes for resource R2 against generalist replicator 3 that can also consume resource R3. Replicator 2 is a specialist with twice the replication ratio of replicator 2 in the similar Selection Model 1.3. Again, the generalist replicator 3 consumes the shared resource R2. (b) The formulas are the same as Model 1.3 except for the replication-ratio increase of replicator 2. (c) The results show what happens when a generalist replicator 3 population competes against a fertile specialist replicator 2. Instead of going extinct, as in Model 1.3, specialist replicator 2 coexists with generalist replicator 3.

values for the shared and unshared resources respectively. If N2=0, the number fed for replicator 3 at the carrying capacity is N3F=R2+R3 and the number resulting from replication is N3=K3(R2+R3). Likewise, if N3=0, the number fed for replicator 2 at the carrying capacity is N2F=R2 and the number resulting from replication is N2=K2 R2. If N2<N3 or equivalently K2<K3E where the equivalent replication ratio K3E is given by

$$K3E=K3(1+R3/R2) \qquad (3.1)$$

then the generalist replicator 3 is growing faster than the specialist replicator 2. Under the Selection Theorem, the faster-growing replicator wins the competition for the shared resource.

Consider now Figure 3.2d and 3.2e that show the resulting specialist and generalist population numbers at time=100 respectively for a number of values of the specialist replication ratio K2 and the unshared resource value R3.

For the conditions of Model 1.3 of K2=K3=2, R2=20, and R3=10, then K3E=3 is greater than K2 which predicts specialist extinction and that is what happens. Note that Figure 3.2d and 3.2e show a threshold at K2=3 as predicted, below which is specialist extinction and above which is specialist coexistence. Figure 3.2d and 3.2e also show the case of R3=20 for which now KE4=4. This is again confirmed as the threshold.

Consequently, there is a critical threshold for the specialist replication ratio of K2=K3E below which there is extinction and above which there is coexistence. The generalist always survives because the non-shared resource is always available to it without competition. But if K2<K3E, then selection occurs against the specialist with respect to the competed resource. This is all consistent with the Selection Theorem. Note how the effective replication ratio K3E varies with the relative size of the resources. Specialist coexistence would not be unexpected in evolution for low N3/N2 but would be increasingly rare for higher values because of the need prior to invasion for a specialist environment that selected for higher replication ratios. Please note that all this applies to the conditions of Figure 3.1a where the shared resource is the preferred choice of the generalist.

What does this all mean for the silver carp and the Bigmouth Buffalo? Clearly if the shared resource is the preferred choice, the unshared resource of the Bigmouth Buffalo generalist is sufficiently small, and the replication ratio of the silver carp sufficiently large, then coexistence occurs. This is exactly what the Phelps et al. (2017) field data shows. Coexistence could also occur if the shared resource is not the preferred choice. The Bigmouth Buffalo population is suppressed but not extinct. However, this does raise the interesting possibility of a different approach to controlling invasive

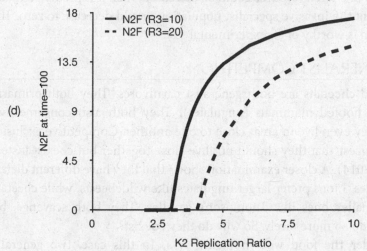

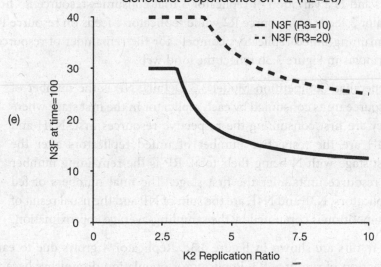

FIGURE 3.2D–E   Invasive Specialist Model 3.2. (d). Specialist population result N2F (time=100) of varying specialist replication ratio K2 and unshared resource R3. The generalist prefers the shared resource R2 and its replication ratio is K3=2. (e). Generalist population result N3F (time=100). Results confirm that specialist extinction occurs if K2<K3(1+R3/R2), otherwise there is coexistence.

specialists. If the shared resource is preferred and somehow the value of the generalist unshared resource could be increased, then it is predicted that at some point the invasive specialist population would decrease to zero. This prediction is worthy of an experimental test.

3.3 GENERALIST COMPETITION

Lions and cheetahs are both generalist carnivores. They both primarily consume hoofed mammals (ungulates). They both also consume live-stock. They even live in areas close to one another. Competitive exclusion might suggest that they should not live close together but coexist instead (Morell, 2014). A closer examination shows that they have different dietary preferences. Lions prefer larger ungulates like wildebeests, while cheetahs prefer smaller ones like Thompson's gazelles. They both scavenge, but cheetahs do so more rarely. So why do they coexist?

Consider the food web in Figure 3.3a. In this case, two generalist replicators consume two different resources first before competing for both resources. Replicator 3 consumes resource R2 first. Replicator 4 appears and is a variation of replicator 3 that consumes resource R3 first. Replicator 3 feeds on resource R2 while replicator 4 feeds on resource R3. The remaining unfed replicators compete for the remainder of resources. The formulas in Figure 3.3b reflect the food web.

> Generalist Competition Model 3.3 Details. NP is the number of resource units consumed by each replicator in the first stage where they are first consuming the respective resources first. N3H and N4H are the respective number of unfed replicators after the first stage with N being their total. RP is the remaining number of resource units after the first stage. The final numbers of fed replicators, N3F and N4F, are the sum of NP and the usual result of competition is computed by the sampling-average approximation.

The results are shown in Figure 3.3c. Replicator 4 grows due to early consumption of resource R3. Replicator 4 population diminishes because its share of R3 diminishes. Replicator 3 does not go extinct and reaches a balance with replicator 4. This model shows that a competition of generalists with different preferences results in coexistence.

So how universal is the coexistence of generalists? Consider the other generalist food webs shown in Figure 3.4. If competing generalists have exactly the same preferences and resources as in Figure 3.4a, it is easy to show that the resources can be collapsed into a single resource and the

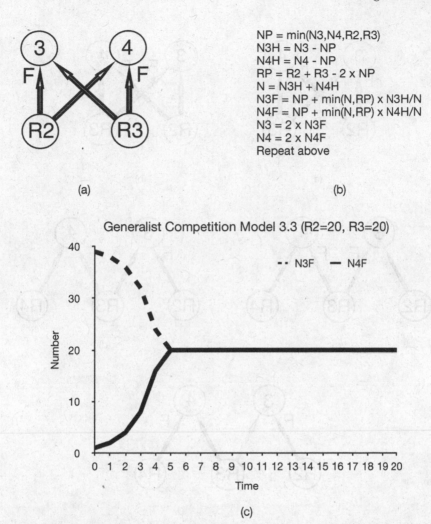

(a)

$$NP = \min(N3,N4,R2,R3)$$
$$N3H = N3 - NP$$
$$N4H = N4 - NP$$
$$RP = R2 + R3 - 2 \times NP$$
$$N = N3H + N4H$$
$$N3F = NP + \min(N,RP) \times N3H/N$$
$$N4F = NP + \min(N,RP) \times N4H/N$$
$$N3 = 2 \times N3F$$
$$N4 = 2 \times N4F$$
Repeat above

(b)

Generalist Competition Model 3.3 (R2=20, R3=20)

(c)

FIGURE 3.3 Generalist Competition Model 3.3. (a) The food web shows a competition between generalist replicators 3 and 4 where each has a separate resource to consume first but then must share them. (b) The formulas reflect the two stages of consumption with NP consumed first by each which is then added to the second stage result of competition with the remainder population and resources. (c) The results show a redistribution due to competition that converges to coexistence.

result is one of the models of Chapter 2 on fertility. Competitive exclusion and the Selection Theorem apply so that one generalist is selected and the other goes extinct. All of the other generalist food webs shown in Figure 3.4b through 3.4e result in coexistence. Coexistence comes easily to generalists. Unless there is identity of resources and preferences, generalists will coexist. It is the opportunity for a generalist replicator to consume a

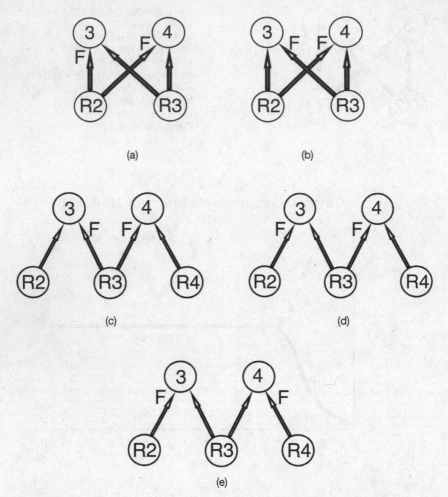

FIGURE 3.4 Other generalist-competition food webs. (a) is equivalent to the simple competition food webs of Chapter 2 for a combined resource R2+R3 and results in selection. (b) is equivalent to Model 3.3 and results in coexistence. All of the rest (c, d, and e) result in coexistence because all replicators have an unshared resource.

resource without competition from other replicator species that enables coexistence.

3.4 COMMENTS AND CONCLUSIONS

In the previous chapter, fertility was king. As long as a replicators had more offspring, they were selected over those with fewer offspring. In an unchanging environment, specialists used streamlining to increase fertility and be selected. But the appearance of new resources changed everything.

In the early stages of life on Earth, it was the vast buildup of detritus that provided these new resources. At first, it was probably the new detritus that was first scavenged because it contained organics that were not far from that used by replicator metabolism. But the vastly larger presence of old detritus offered the potential for greater growth and thereby selection. It would be a relatively small evolutionary step to go from scavenging new detritus to scavenging slightly older detritus and the process would have continued.

Variations appeared in replicators to try consuming the new resources while still having a preference for the old resource. These were the first generalists. This variation was relatively small and so was consistent with a gradual evolutionary change. To start with the old resource and suddenly jump to a preference for the new resource was a much larger change and so was relatively unlikely at first. To have a preference for the new resource, it must build up gradually. This must have occurred by starting with a preference for the old resource and then gradually evolving an increased use of the new resource until it became the preferred resource. This would have naturally occurred if the new resource had advantages such as greater nutritional value or availability. The switch to preference for the new resource would have been built up over evolutionary time. In the early life on Earth, the new resource value of detritus would have been the greater availability of older detritus.

So now there is a world of specialists and generalists scavenging on detritus. But when they interact, it is hard to be a specialist. If a new resource appears amidst a population of specialists, the initial replicator variation still has a preference for the old resource. As illustrated in Figure 3.1, the result is that the old specialist replicators go extinct and the new generalist replicators are selected. If new resources do not appear for a while, then other specialists have a chance to survive the generalist interaction. If the generalists evolve to have a preference for the new resource, then the other specialists can coexist with them. Also, if the new resource is relatively small, then other specialists could evolve sufficiently greater fertility to coexist with the generalist interaction. Otherwise, the specialists lose the interaction with the generalists.

The generalists have a much easier time. With one exception, generalists easily coexist with specialists and other generalists. Only when other generalists have exactly the same resources and preferences does interaction result in extinction and selection. If the concept of competitive exclusion is updated to include resource preference, then all of this is consistent with it. Resource preference is a very powerful force in evolution and including it as part of the conditions for competitive exclusion resolves much of the ambiguity.

Thus, the appearance of more complex environments with interaction between specialists and generalists is the first step in the evolution of complexity. Initial replicators change the ecology by creating detritus which provides resource variation. In turn, the replicators evolved more complexity in response to these resource variations. Some specialist replicators will find simple environments and evolve simpler capabilities to best adapt to the simpler environments. But other generalist replicators will evolve more complex capabilities to adapt to more complex environments. Figure 3.5 illustrates this idea which is named "The Tao of Replicators." Replicators in simpler environments will evolve to become simpler replicators while replicators in more complex environments will evolve to become more complex replicators.

Finally, the replicator rules for specialist and generalist interaction that have been described in this chapter do not just apply to early scavengers but to all of evolutionary history. The following chapters will show how carnivores are governed by these replicator rules. But it is instructive to realize that even human evolutionary history was impacted by them. The Neanderthals were almost exclusively carnivores. Analysis of Neanderthal bone marrow (Jaouen et al. 2019) supports this and so they were specialists. Early homo sapiens consumed a variety of food but with a preference for

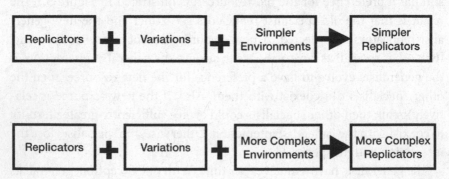

FIGURE 3.5   The Tao of Replicators. Evolution is about adaptation to either simpler or more complex environments with the corresponding results.

meat (Cordain et al. 2002). So they were omnivores and generalists with a preference for a resource shared with the specialist Neanderthals. The prediction of replicator specialist-generalist interaction is that the Neanderthals would go extinct and that the homo sapiens would be selected. And that is exactly what happened.

CHAPTER 4

Predators

I T WAS A SMALL variation from consuming fresh detritus to consuming live prey. But predators took that step and evolution was never the same. One small step for a microbe, one giant leap for complexity. In a single stroke, predators changed everything. Predators wiped out countless species. For example, in the last few hundred years, the introduction of foxes and feral cats into the ecosystem of Australia caused the extinction of numerous mammals and birds (Dickman, 1996). Predators forced evolutionary change on prey in order to avoid extinction. In turn, the evolution of predator-resistant prey forced predators to adapt. This was the interactive ecological change that drove the evolution of novelty and complexity.

Without predators, evolution would have taken a totally different path. As Bengston (2002) points out:

> An imaginary biosphere without predators would be very different from what we are acquainted with. There would be various kinds of photo- and chemoautotrophs making use of available energy gradients to reduce carbon for energy storage and constructional/ physiological purposes. There would be organisms scavenging excess organic matter, but there would be no organisms directly interrupting the lives of others by pilfering their tissues...the selective pressures would be very different from those that affect

DOI: 10.1201/9781003391395-4

most organisms today. Survival requirements would center around positioning oneself with respect to chemical, temperature, and light gradients, and the only need to move would be in order to adopt shifting gradients—for example, varying light intensities or redox boundaries. Although competitive interactions would not be excluded, they would mostly be related to relative efficiencies of energy conversion systems. Under such circumstances diversities would be low and stable. The most complex benthic ecosystems would likely be layers of physiologically differentiated microbes, i.e., microbial mats.

The world with predators evolved in a totally different way and is reflected in its much greater diversity and complexity. Although predators have existed for a very long time, early predators faced a number of evolutionary challenges that they had to overcome or coexist with. Predators changed the ecology around them. But the ecology also had to change to support them. Such issues include overfeeding, cannibalism, prey defense, generalization, and cyclic behavior. This chapter will examine each of these challenges in turn.

Our models now also evolve from just replicator models to more complex models. In order to meet the challenges that predators and prey face, variations capable of growing more complex features are needed. The addition of these features makes the models more like complex life than just replicators that only consume and replicate. This chapter shows how introducing these features solves certain critical problems and the next chapter shows how they can grow to achieve the evolution of novelty and complexity.

4.1 OVERFEEDING

Georgyi Gause (1934) also published the results of another microbe experiment that this time involved predators and prey. As a prey, he again used *Paramecium* (*P. caudatum*) in a medium of nutrients and water. As the predator, Gause added *Didinium* (*D. nasutum*). Their preferred prey is *Paramecium*. This predator microbe is voracious and uses filaments carrying a poison to ensnare and paralyze its prey. When Gause introduced the predator to the culture of prey, the predator devoured them all. Then lacking a food supply, the predator starved and expired.

Overfeeding is one of the greatest dangers for predators. A predator population may gorge themselves on a defenseless prey population and wipe them out, but then may face deprivation because of the absence of a food resource. For example, human commercial overfishing is seriously risking the extinction of many marine species. The bluefin tuna has effectively vanished from the North Sea and is in severe decline in the western Mediterranean Sea. Such overfishing has led to the massive destabilization of marine ecosystems (Hogan, 2014).

Consider the food web in Figure 4.1a. Resource R3 consists of fresh detritus which is consumed by prey 4. Predator 5 is a variation that consumes prey 4. Figure 4.1b shows the formulas whereby prey 4 eats first

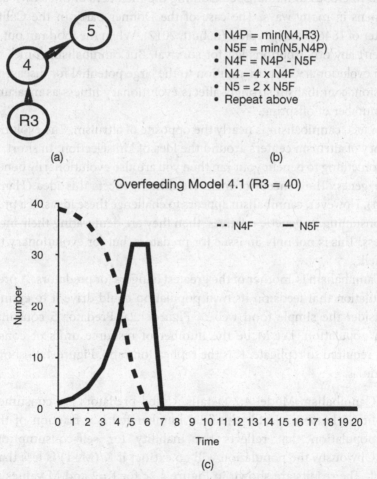

- N4P = min(N4,R3)
- N5F = min(N5,N4P)
- N4F = N4P - N5F
- N4 = 4 x N4F
- N5 = 2 x N5F
- Repeat above

(a)

(b)

(c)

FIGURE 4.1   Overfeeding Model 4.1. (a) The food web shows a predator 5, a prey 4, and a resource R3 for the prey. (b) The formulas show the prey feeding first followed by the predators. (c) The results demonstrate the effects of overfeeding and the consequent extinction for both prey and predators.

and then predator 5 preys on it. Note that prey 4 is more fertile (4X) than predator 5 (2X), a common occurrence among prey and predators.

Figure 4.1c shows the results. At first, the predator 5 population grows and the prey 4 population declines. Eventually, prey 4 goes extinct. Because the resource for predator 5 disappears, it too then goes extinct. This model demonstrates the danger of overfeeding.

4.2 CANNIBALISM

Cannibalism is common among predator species (Elgar and Crespi, 1992). From microbes to humans, consuming the members of one's own species happens in many ways. The case of the Donner party in the California winter of 1846 is fairly typical (Schutt, 2017). When the food ran out, there weren't any other options left for survival. But cannibalism comes with a high evolutionary cost. In addition to the large potential for disease transmission, cannibalism strongly affects evolutionary fitness as measured by the number of offspring.

In fact, cannibalism is nearly the opposite of altruism. The evolutionary theory of altruism centers around the idea of kin selection. In short, if you do something to benefit your kin then you are also evolutionarily benefiting your genes. The concept of inclusive fitness reflects this idea (Hamilton, 1964). However, cannibalism appears to challenge these ideas. If a predator is consuming its genetic relatives, then they are denigrating their inclusive fitness. This is not only an issue for predators, but for evolutionary theory as well.

Cannibalism is another of the greatest dangers for predators. A predator population that feeds on its own population could drive it to extinction. Consider the simple food web of Figure 4.2a. Predator 5 consumes its own population. Let M be the number of resource units of consumption required to replicate. K is the replication ratio. Figure 4.2b shows the formulas.

> Cannibalism Model 4.2 Details. Unfed predators are consumed first. The formula for the final number fed is a fraction of the population that reflects the inability for self-consumption. Obviously, the population will go extinct if $K/(M+1)$ is less than 1. The results are shown in Figure 4.2c for K=2 and M values of 1.5, 2, and 3.

It is clear that sustained exclusive cannibalism of in-group predators ends in extinction. It is judged that $K>M+1$ is unlikely due to insufficient

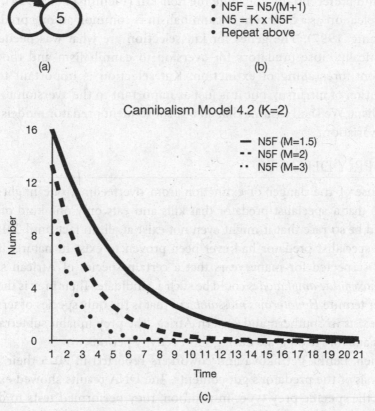

(a)

- N5F = N5/(M+1)
- N5 = K x N5F
- Repeat above

(b)

Cannibalism Model 4.2 (K=2)

— N5F (M=1.5)
-- N5F (M=2)
·· N5F (M=3)

(c)

FIGURE 4.2 Cannibalism Model 4.2. (a) The food web shows self consumption of the population. (b) The formulas assume consumption of M units for replication with a ratio of K. (c) The results end in extinction for K=2 and M equal to 1.5, 2, and 3.

resources for sustainment and replication. However, occasional or out-group cannibalism is not prohibited and does exist in nature (Elgar and Crespi, 1992). Nevertheless, a fundamental variation that is inherited by all successful predators must appear that has a preference aversion to sustained exclusive in-group cannibalism. Such a variation would deter extinction of predators from cannibalism.

Fortunately, evolution has provided a remedy. Consider the case of tiger salamanders which are widespread across the United States. When their eggs hatch, the larvae are similar to tadpoles. These larvae normally feed on zooplankton or very small invertebrates. Although under crowded conditions they become cannibals, they also discriminate between kin and non-kin. Through their sense of smell, they recognize kin and preferentially consume the non-kin (Pfennig et al., 1994). This kin selection as a deterrent to cannibalism is common among predators (Pfennig, 1997). The genes for kin selection are what was needed by genetically close predators for aversion to cannibalism and the subsequent forestalling of extinction. Kin selection is important to the evolution of altruism, but it is just as important to the aversion to cannibalism. We shall assume that all our subsequent predator models have this variation.

4.3 PREY DEFENSE

Because of the danger of extinction from overfeeding, one might conclude that a specialist predator that kills and eats only one kind of prey would be so rare that it might even not exist at all. In fact until 2015, no such specialist predator had ever been proven to exist in nature. It had been suspected for many years that a certain species of African spider *Ammoxenidae amphalodes* could be such a candidate. Their prey is the harvester termite *Hodotermes mossambicus* that is the only species of termites that exists in southern and eastern Africa. The prey inhabit subterranean nests and forage on grass during short activity periods.

Then, Lenka Petrakova and coworkers reported in 2015 their DNA analysis of the predator's gut contents. The DNA results showed exactly only the specific prey type. In addition, they performed tests to determine the willingness of the predator to accept other similar prey types. The acceptance tests showed that the predator totally ignored the other similar prey types. These results confirmed that this predator is the first proven example of a predator that kills and eats only one kind of prey. So how does this predator overcome the danger of extinction from overfeeding?

As long as prey have no defense against predators, the predators will drive the prey to extinction. And if the predators have no other resources, they too will go extinct. There are numerous potential defense mechanisms for prey that could allow them to escape predators, everything from using the natural environment for concealment to evolving better escape or defense features. In turn, the evolution of prey defense can lessen the chance for predator extinction.

Figure 4.3a shows the food web for a model that begins to explore prey defense. In this model, a new variation of the prey 4 of Model 4.1 appears. This new prey has a countermeasure that limits predation. On the food web, this is denoted by the letter "C".

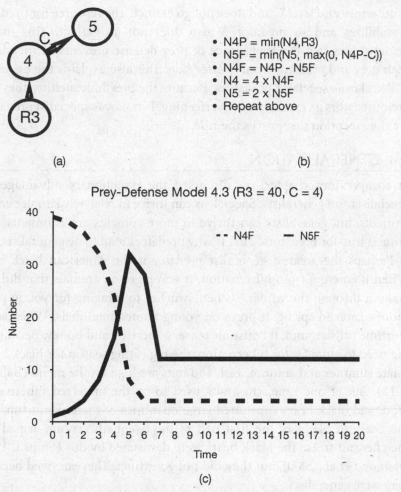

- N4P = min(N4,R3)
- N5F = min(N5, max(0, N4P-C))
- N4F = N4P - N5F
- N4 = 4 x N4F
- N5 = 2 x N5F
- Repeat above

(a)

(b)

Prey-Defense Model 4.3 (R3 = 40, C = 4)

-- N4F   — N5F

(c)

FIGURE 4.3   Prey-Defense Model 4.3. The food web (a) and the formulas (b) are the same as Model 4.1 except for the countermeasure "C" that limits predation. (c) The results show that prey defense can deter extinction from overfeeding for both prey and predators.

Prey-Defense Model 4.3 Details. As before in Model 4.1, prey 4 consumes resource R3 and the predator 5 consumes prey 4. The formulas are shown in Figure 4.3b. They are nearly the same as the ones for Model 4.1. The difference is that the amount of prey consumption by the predator is limited by the countermeasure constant C.

The results are shown in Figure 4.3c. As before, predator 5 appears and its population grows. Prey 4 declines. Eventually, prey 4 reaches countermeasure level C and does not go extinct. The resource for predator 5 stabilizes and so predator 5 also does not go extinct. This model demonstrates that the appearance of prey defense prevents extinction of both prey and predator from overfeeding. This also explains the existence of Petrakova's specialist predators. Because the prey hide, neither they nor their predators go extinct from overfeeding. Petrakova's specialist predators are the exception that proves the rule.

4.4 GENERALIZATION

In the previous chapter, we discussed the evolutionary advantages of specialists and generalists. Specialists can thrive in relatively simpler environments, but generalists can thrive in more complex environments. The same is true for predators. That is why predators tend to be generalists.

Perhaps the greatest generalist predator is the American black bear. When it emerges from hibernation, it scavenges on animals that did not make it through the winter. It then switches to foraging for young plant shoots. Later in spring, it preys on young hoofed mammals. When summertime rolls around, it consumes insects, berries, and honey. Because of the need to store fat for hibernation, the big eating season for black bears is late summer and autumn. Fish and nuts are high on the menu (Ballard, 2013). But at one time, chestnuts used to be the preferred hibernation food, and black bears consumed large quantities of them. Unfortunately, the chestnut blight of the first half of the twentieth century wiped out the chestnut trees. The black bears were devastated by this (Stupka, 1960; Diamond et al., 2000), but they did not go extinct. They survived because they were generalists.

Predators are frequently generalists and consume more than one kind of prey (e. g., Hayward et al., 2006). Figure 4.4a shows a food web that adds an additional resource R4 for predator 5 to the food web of Model 4.3. As indicated by the arrow with the letter "F," predator 5 consumes the prey 4 first and then consumes the new resource R4. Resource R4 models a second kind of prey as a resource flow.

> Generalization Model 4.4 Details. The formulas of Figure 4.4b reflect the multi-stage approach to computing consumption from multiple resources. First, the prey eats. Then the predator-prey consumption takes place. Finally, the unfed predators consume resource R4 and the result is totaled with the predator consumption of the prey.

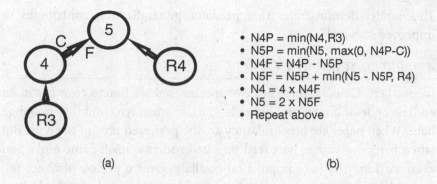

* N4P = min(N4,R3)
* N5P = min(N5, max(0, N4P-C))
* N4F = N4P - N5P
* N5F = N5P + min(N5 - N5P, R4)
* N4 = 4 x N4F
* N5 = 2 x N5F
* Repeat above

(a)   (b)

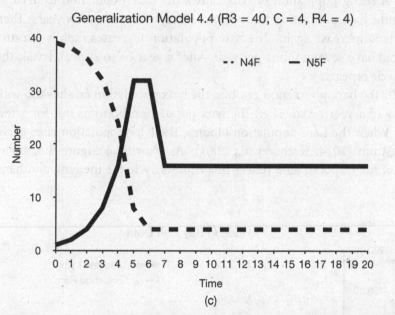

(c)

FIGURE 4.4   Generalization Model 4.4. (a) The food web shows predator 5 generalizing its resources over Model 4.3 with the addition of resource R4. (b) The formulas reflect the multi-stage computation with prey consumption, then predator consumption of prey first as indicated by the arrow with the letter "F," then finally unfed predator consumption of resource R4 and the result totaled. (c) The results show that predator generalization contributes to an increase in final predator population over resource specialization like that in Model 4.3.

The results are shown in Figure 4.4c. Predator 5 population grows while prey 4 population declines. Eventually, prey 4 reaches countermeasure level C and does not go extinct. Because the prey resource stabilizes and there is an additional resource, predator 5 population also stabilizes, this time to a level appropriate to the sum of resources.

This model demonstrates that predator generalization contributes to improved survival.

4.5 PREDATOR-PREY CYCLES

In northern Canada, there are two species that are bound together in an endless cycle of hunter and hunted—the Canadian lynx and the snowshoe hare. When hares are plentiful, they are the preferred prey of the lynx. But when hares are scarce, lynx feed on small rodents, small game birds, and even carrion. The hare population oscillates over a period of about ten years. A rising population of lynx causes the hare population to decline. When the hare population hits bottom, it stays there for a few years. Then it starts to increase again. The hare population increases rapidly because they can have several litters per year. After a year or so at peak levels, the hare cycle repeats.

After the hare population crashes, the lynx population crashes too with a delay of a year or so. Then the lynx population bottoms out for a few years. When the hare population blooms, the lynx population rises again (Hoagstrom, 2014; Krebs et al., 2001). As shown in Figure 4.5, many years of fur-trapping data reflect this endless cycle for the lynx and hare.

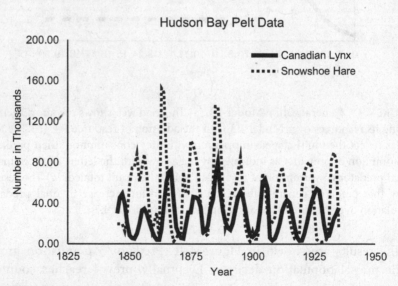

FIGURE 4.5 Historical Hudson Bay Company fur-trapping data for Canadian lynx and snowshoe hare by year. This data is the original evidence of predator-prey cycles (from Hewitt, 1921).

Although this cyclic behavior is relatively rare, other species have shown it too (Hörnfeldt, 1978).

It would be expected that eventually a predator variation would appear that has a variable resource preference that depends on abundance. This would clearly have an advantage over a fixed preference for an ecology with varying resources such as one for predators and prey. However, a variable preference would require some kind of switch that flips one way for resource abundance and the other way for resource scarcity. Fortunately, switches are ubiquitous in biological organisms. They are the workhorses of genetic regulation, evo-devo, and metabolism control (e.g., Carroll, 2005).

Consider the food web shown in Figure 4.6a. It is almost the same as that for Model 4.4. But this time the predator 5 has a switch denoted by the letter "S." S is a switch that selects consumption of prey 4 if it is abundant. Switch S selects consumption of resource R4 if prey 4 is not abundant. Finally, switch S is unchanged if abundance change is not detected.

> Predator-Prey-Cycles Model 4.5 Details. The formulas shown in Figure 4.6b are a modification of the formulas in Figure 4.4b that now add the switch S. G1 is a gate that detects if the prey 4 population N4 is above the threshold TU. G2 is another gate that detects if N4 is below the threshold TL. G3 is a gate that detects an "on" state for either gate G1 or gate G2. S is the switch that takes the value of G1 if G3 is "on" and remains unchanged otherwise. Consumption of prey 4 or the resource R4 then follows depending on the state of the switch S.

The results are shown in Figure 4.6c. Prey 4 population initially declines as predator 5 population rises. As the prey 4 population bottoms out, the predator 5 population crashes. Because predator 5 has switched to consuming resource R4, prey 4 population blooms. Predator 5 then switches to consuming prey 4 and the cycle repeats. Once again, this model demonstrates the evolutionary power of resource preference.

4.6 COMMENTS AND CONCLUSIONS

About 541 million years ago, the beginning of the Cambrian era marked the fossil record from having relatively few fossils to having very much more. This became known as the Cambrian explosion. The Cambrian fossils contained skeletons of the most amazing creatures ever to exist on Earth.

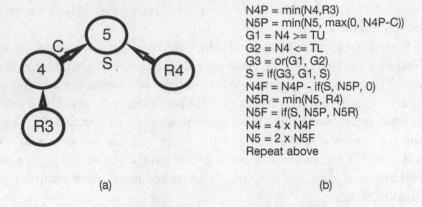

N4P = min(N4,R3)
N5P = min(N5, max(0, N4P-C))
G1 = N4 >= TU
G2 = N4 <= TL
G3 = or(G1, G2)
S = if(G3, G1, S)
N4F = N4P - if(S, N5P, 0)
N5R = min(N5, R4)
N5F = if(S, N5P, N5R)
N4 = 4 x N4F
N5 = 2 x N5F
Repeat above

(a)   (b)

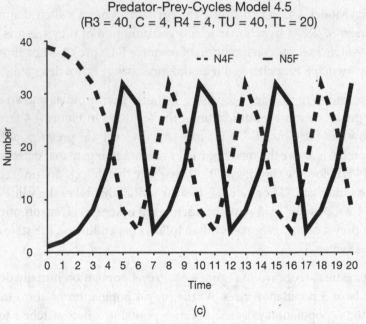

Predator-Prey-Cycles Model 4.5
(R3 = 40, C = 4, R4 = 4, TU = 40, TL = 20)

(c)

FIGURE 4.6 Predator-Prey-Cycles Model 4.5. (a) The food web adds a switch "S" to predator 5 of the food web in Figure 4.4a. (b) The formulas build on those of Figure 4.4b with the addition of the switch S. Switch S is made up of gate G1 that detects prey 4 abundance, gate G2 that detects prey 4 scarcity, and gate G3 that detects either. Switch S is unchanged unless G3 is on. If S is on, prey 4 is consumed. If S is off, resource R4 is consumed. (c) The results show a predator bloom, a prey crash, a predator crash, a prey bloom, and a cycle repeat. This model demonstrates the basic predator-prey cyclic behavior.

Armored, fanged, clawed, and multi-legged monsters that were beyond human imagination (Gould, 1989). This was an era when predators played a strong role as indicated by many fossils that bore the marks of predators.

But predators must have existed before the Cambrian era. Just before it, a curious creature known as Cloudina existed. The creature was about one to ten inches long and consisted of nested cones. It was a little like a snail with a shell that had no chambers, but was relatively straight rather than spiraled. The interesting thing about Cloudina is that their fossils contained puncture-like holes that indicated the first direct evidence of predators (Bengston and Zhao, 1992).

Yet even still, predators must have existed long before Cloudina. Bengston (2002) lays out the evidence for this, but the relative lack of direct predator fossils limits the lines of evidence. Possibly, the most compelling argument for predators long before the Cambrian era is the modern existence of microbe predators. If modern microbes have predators, then there is every reason to believe that relatively early in the history of life, in the teeming microbe world billions of years ago, there must have been predator microbes too.

But this begs the question—if predators played such a strong role in the Cambrian explosion and they existed long before that, then why did it take so long for there to be an explosion of ecological diversity like that found in the Cambrian era? There are many ideas that could contribute to providing solutions to this puzzle such as the evolution of hard bodies and the HOX genes for replicating body parts. But this chapter has provided a different set of clues. That is that predators had a number of evolutionary challenges that required solutions before predators could drive more complex evolution. Those challenges include overfeeding, cannibalism, prey defense, generalization, and cyclic behavior.

Overfeeding was a bane of predators. It could wipe out species wholesale and when they were gone, the predators would starve and go extinct too. Predator generalization helps, but in some sense, it just delayed the inevitable. If predators wipe out all multiple sources of prey, then extinction still follows. Prey defense was the better resolution because that prevented both predator and prey extinction. But that meant that the survival of predators depended on the evolution of prey and that could have taken a while. There was cannibalism which could be species suicide. To prevent that, evolution of kin selection was required to enable aversion to cannibalism. Finally, there was the boom and bust of predator-prey cycles that brought both dangerously close to extinction with every crash. However, predator-prey cycles are a preview and an echo of arms races.

With so many difficult evolutionary challenges, it is then no wonder that predators took so long to reach the Cambrian explosion. But predators did meet these challenges and when they did, they were ready for the evolutionary explosion in novelty, complexity, and diversity that is discussed in the next chapter.

CHAPTER 5

Arms Races

COEVOLUTION IS ALL AROUND us. The flowers evolve more beautiful blooms, and the bees are attracted to them. The bees evolve better navigation skills and then they can find the flowers better and return to the hive. And the cycle repeats. The rabbits evolve greater speed, and they don't end up as lunch. The foxes also evolve faster sprints and then they can eat lunch. Even amongst humans, coevolution is everywhere. Highway speed traps use increasingly sophisticated radar and lidar to catch speeders. Better and better radar and lidar detectors show up to counter such traps. Even within our bodies, coevolution is going on right now. Antibodies are created to fight infection, while pathogens evolve to defeat antibodies. Drugs are developed to help fight pathogens and they in turn evolve to become drug resistant. It's a coevolving world. And a particularly vigorous kind of coevolution is the arms race.

Evolutionary arms races have long been recognized as one of the major drivers of evolution (Dawkins, 1986). A typical arms race is where one side of a conflict gains an advantage and the other side counters, gains the advantage, and the cycle repeats. The US-USSR ballistic-missile escalation is one historical example.

Even Darwin noted the constant evolutionary struggle between predators and prey. For example, a prey evolves to run faster to escape predators, the

DOI: 10.1201/9781003391395-5

predators evolve to run even faster to be able to catch the prey, and the cycle repeats many times. Numerous examples in biology have been identified with several different kinds of arms races (Dawkins and Krebs, 1979). But how exactly does this work in evolution? In this chapter, we shall explore arms races between predators and prey. Chapter 7 will discuss arms races between infectious diseases and hosts.

One kind of arms race is an apparently minimal amount of change in the relationship between the predators and prey. The predators still hunt, and the prey are still hunted even though both are adapting to gain advantages. To describe this phenomenon, the "Red Queen's Hypothesis" was proposed (Van Valen, 1973) that organisms had to adapt as quickly as possible just to stay in place. The name is taken from Lewis Carroll's book *Through the Looking Glass* where Alice was walking and then running with the Red Queen and complained that she was running as fast as she could but not getting anywhere. The Red Queen replied, "it takes all the running you can do, to keep in the same place."

In this chapter, we shall demonstrate that models can successfully illuminate arms races and the Red Queen's Hypothesis. We shall develop a base arms-race model and then exercise it for several cases. The result will be ever more novel and complex creatures and ecologies. With the advent of arms races, the models enter a complex-ecologies phase.

5.1 BASE ARMS-RACE MODEL

The food web of the Base Arms-Race Model (BARM) is shown in Figure 5.1a. BARM is a combination of Selection Model 1.3 and the Prey-Defense Model 4.3. 5 and 7 are predators that share prey 4 and 6 which in turn share resource R3. Every predator-prey consumption arrow gets a different countermeasure constant C. Prey 4 is consumed first. Variations will consist of incrementing C values for new prey and decrementing C values for new predators. Variants of BARM will be employed to illustrate different aspects of arms races.

The formulas for BARM are shown in Figure 5.1b.

> BARM Details. Again, we employ a multi-stage approach to computation. The functions f and g are defined for formula compactness. Then the usual resource sharing for prey 4 and 6 is computed to give the preliminary numbers N4P and N6P. Next, resource sharing of prey 4 for the predators 5 and 7 starts with computing the resource amount RF that is available to both. That amount is

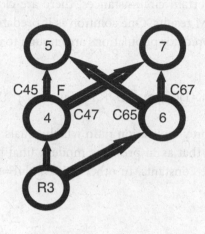

f(x,y)=min(x,y)/x
g(x,y)=max(0,x-y)
N=N4+N6
N4P=N4 f(N,R3),
N6P=N6 f(N,R3)
RF=min(g(N4P,C45),g(N4P,C47))
NCF=N5+N7
N5P=N5 f(NCF,RF)
N7P=N7 f(NCF,RF)
N4T=N4P-N5P-N7P
N5FP=N5P+min(N5-N5P, g(N4T,C45))
N7FP=N7P+min(N7-N7P, g(N4T,C47))
N4F=N4P-N5FP-N7FP
RG=min(g(N6P,C65), g(N6P,C67))
N5H=N5-N5FP
N7H=N7-N7FP
NCG=N5H+N7H
N5Q=N5H f(NCG,RG)
N7Q=N7H f(NCG,RG)
N6T=N6P-N5Q-N7Q
N5T=N5Q+min(N5H-N5Q, g(N6T,C65))
N7T=N7Q+min(N7H-N7Q, g(N6T,C67))
N6F=N6P-N5T-N7T
N5F=N5FP+N5T
N7F=N7FP+N7T
N4=4xN4F
N5=2xN5F
N6=4xN6F
N7=2xN7F
Repeat Above

(a)   (b)

FIGURE 5.1   Base Arms-Race Model (BARM). (a) The food web shows two predators 5 and 7 sharing two prey with countermeasure constants, first 4 then 6, which share resource R3. (b) The formulas are based on those for Models 1.3 and 4.3. The functions f and g are for compactness. First prey 4 and 6 share R3. Then, predators 5 and 7 share prey 4. Next, unfed predators 5 and 7 share prey 6. Finally, the results are totaled and replication takes place.

shared and the remaining amount of prey 4, N4T, is computed. Then the unfed predators for 5 and 7 are fed and totaled with the shared amounts to give N5FP and N7FP. These are used with the prey 4 preliminary number N4P to compute the final number N4F. The same process of predator consumption of prey 4 is repeated for prey 6 and for the unfed predators and the results are appropriately totaled.

This model with the shown numbering is designated

$$(N4,N5,N6,N7)=BARM(N4,N5,N6,N7) \qquad (5.1)$$

This means that we are employing the original model numbering system. In the following cases, we shall renumber as appropriate as evolution progresses, and this will be denoted by changing the labels of the left side of this equation.

It should be noted that under certain circumstances, there are closed analytic solutions to the final BARM results. One solution is if predator 5 goes extinct (N5F=0) and if initial predator populations are enough to prevent premature termination, then

$$N7F = if(R3>K\ CT, (K-1)CT, max(0, R3-CT)) \tag{5.2}$$

where CT=C47+C67 and K is the prey replication ratio which equals 4 in Figure 5.1b. Also note in this case that as in previous models, final prey numbers equal the countermeasure constants. In other words, N4F=C47 and N6F=C67.

5.2 PREDATOR VARIATION

When predators first appeared, prey had to evolve to defend themselves. Their defenses took many forms from hiding to weapons. But for prey that were not very mobile or strong, the choice of last resort was the use of poisons. At minimum, the predators would find that the prey tasted bad and avoided it. But if that didn't work, the prey would evolve stronger poisons that could stun or even kill. But sometimes, predator variations would appear that countered the poisons. Then the prey evolved even stronger poisons with the predators evolving as well and evolution went a little haywire.

Such a case of evolution running amok happened in the western coastal mountains of North America between the local newts and snakes (Brodie, 2010; Reimche et al., 2020). Apparently, the newts took exception to being a favorite lunch for the snakes and managed to evolve the capability for exuding a deadly poison known as TTX (tetrodotoxin) when attacked. But the joke was on the newts because an ancient ancestor of the snakes had already evolved a partial resistance to TTX (McGlothlin et al., 2016). So, if at first you don't succeed, try, try again. That's what the newts did and evolved to exude even larger amounts of TTX. But this time the snakes were able to adapt and still dine on newts. This vicious cycle was repeated until newts were so toxic that one of them could kill an entire roomful of people. But the snakes had the last laugh and evolved unbelievably high resistance to TTX and continued with their newt munching.

Consider the food-web timeframes for the first BARM case as shown in Figure 5.2a. In the first frame, we start out at the conclusion of Prey-Defense Model 4.3 with predator 5, prey 4, but this time with a counter-measure constant C45=12. Then in the next frame, a predator 6 appears as a predator variation that gives an advantage by decreasing the counter-measure constant to C46=8. BARM is then exercised with the formulas shown in Figure 5.2b.

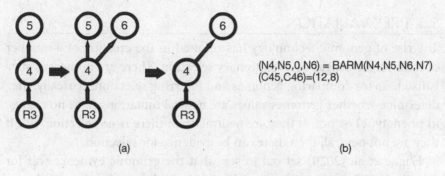

(N4,N5,0,N6) = BARM(N4,N5,N6,N7)
(C45,C46)=(12,8)

(a)   (b)

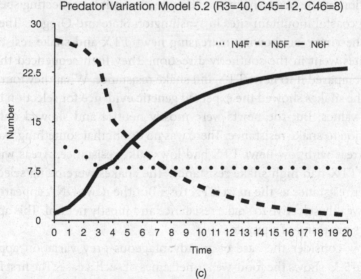

(c)

FIGURE 5.2 Predator Variation Model 5.2. (a) The food-web timeframes start with Prey-Defense Model 4.2, then a predator variation 6 appears with lower C values, and the final resulting frame results with predator 5 extinction. (b) The formulas show the values selected for BARM and the parameters. (c) The results show the rise of predator 6, the extinction of predator 5, and the adjustment to a new countermeasure value for prey 4. This model demonstrates how predator advantages win.

The result is shown in Figure 5.2c. Similarly to Selection Model 1.3, predator 6 has an extra resource to consume because of the difference in C values. Predator 6 population grows. Predator 5 population declines because of a decreasing share of the prey 4 population. Prey 4 adapts to the C value with predator 6. Predator 5 goes extinct as shown in the last frame of Figure 5.2a. This model demonstrates that predators with offensive advantages (i.e., decreased countermeasure effects) will win the competition against less capable predators.

5.3 PREY VARIATION

The rise of genomic technology has resulted in the creation of a number of tools for investigating evolutionary selection. There are numerous statistical tests for comparing genomes and inferring selection. Basically, they determine whether genomes values are neutral mutations (i. e., no change in phenotype) or not. If they are neutral, then there is no selection. But if they are not neutral, then there can be evidence for selection.

Hague et al. (2020) set out to see what the genomic evidence was for selection in the newt-snake war. They started out by collecting specimens along coastal mountain sites in Washington State and Oregon. These sites had the virtue of showing increasing newt TTX and snake resistance as the sites went in the southerly direction. They then sequenced the DNA and compared it to newt TTX and snake resistance. What they found what that the snakes showed the expected genetic evidence for selection by newt TTX values. But the newts were mostly neutral and showed only weak selection to snake resistance. There was no doubt that something was going on. Areas with low newt TTX had low snake resistance. Areas with high newt TTX had high snake resistance. The snakes were being selected for higher resistance as the newt TTX rose. But the newt DNA appeared to be only weakly selective to snake resistance and mostly neutral. This appeared to be something of a puzzle.

Now consider the case of an advantageous prey variation appearing. Figure 5.3a shows the food-web timeframes of such a case. The first frame is the last frame of Predator Variation Model 5.2 with prey 4, predator 6, and countermeasure constant C46=8. In the next frame, prey gain an advantage with the appearance of prey 7 that is a variation which increases the countermeasure value to C76=12. Figure 5.3b shows the BARM naming convention and parameter values.

Figure 5.3c and the last frame of Figure 5.3a show the results. Prey 7 population rises to its C value of C76=12. Prey 4 population stays the same

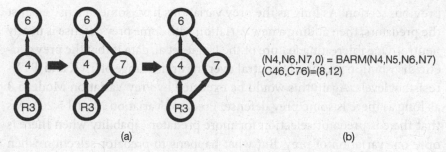

(a)   (b)

$(N4,N6,N7,0) = BARM(N4,N5,N6,N7)$
$(C46,C76)=(8,12)$

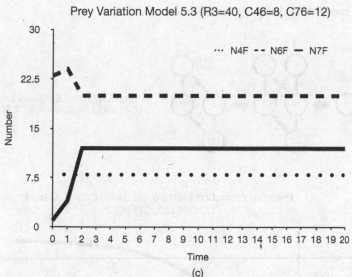

Prey Variation Model 5.3 (R3=40, C46=8, C76=12)

(c)

FIGURE 5.3   Prey Variation Model 5.3. (a) The food-web timeframes show the start at the end of Model 5.2, the appearance of a prey variation 7 with an increased countermeasure value, and the result. (b) The formulas show the BARM naming convention and parameter values. (c) The results show that the prey population rises to its C value, none go extinct, and all reach a coexistence consistent with the BARM analytic solutions. This model demonstrates that extinction and selection are not always consequences of prey advantage.

at its C value. Because of increased C for prey 7, available prey for predators are down. Predator 6 population decreases, consistent with the BARM analytic solutions. No predators or prey go extinct. This model demonstrates that prey advantage does not necessarily cause extinction and selection.

5.4 PREDATOR EXTINCTION

The Hague et al. (2020) data showed that the prey distribution was nearly neutral. This is consistent with Prey Variation Model 5.3 presented in the previous section. As long as the prey variations have some defense against the predators, then adding a new variation with some prey defense is nearly neutral. One interesting feature of the Hague et al. data is that the prey genetic distribution was nearly neutral over the entire range of TTX and TTX-resistant levels. Again this would be expected by Prey Variation Model 5.3 as long as there is some prey defense. Predator Variation Model 5.2 shows that there is predator selection for more predator capability when there is only one variation of prey. But what happens to predator selection when there is more than one prey variation?

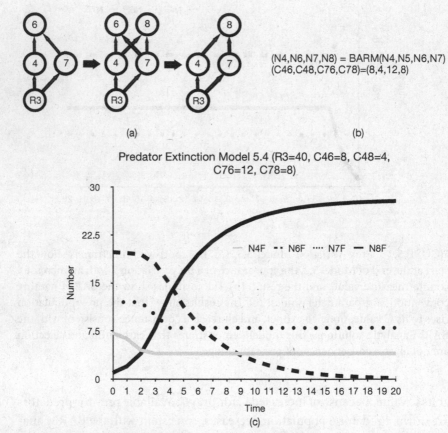

(a)

(N4,N6,N7,N8) = BARM(N4,N5,N6,N7)
(C46,C48,C76,C78)=(8,4,12,8)

(b)

Predator Extinction Model 5.4 (R3=40, C46=8, C48=4, C76=12, C78=8)

(c)

FIGURE 5.4   Predator Extinction Model 5.4. (a) The food-web timeframes show the start with the end of Model 5.3, the appearance of variation predator 8 with countermeasure advantages, and the final result. (b) The formulas show BARM naming and parameter values. (c) The results show the rise of the predator 8 population, the adaptation of the prey to its countermeasure values, and the extinction of predator 6. This model demonstrates that despite multiple prey, predator variations can force ancestor extinction.

As shown in the first timeframe of Figure 5.4a, this model starts at the ending of Model 5.3 with prey 4 and 7 and predator 6 with countermeasure constants C46=8 and C76=12. Now predator variation 8 appears with the advantage of countermeasure decreases to C48=4 and C78=8. Figure 5.4b shows the naming for BARM and the parameter values.

Figure 5.4c and the last timeframe of Fig. 5.4a show the results. Similarly to Predator Variation Model 5.2, predator 8 has an extra resource to consume because of differences in C values. Predator 8 population grows. Predator 6 population declines because of a decreasing share of the prey 4 and 7 populations. Prey 4 and 7 adapt to C values with predator 8. Predator 6 goes extinct. This model demonstrates that even with multiple prey populations, new predator advantages can force ancestor predators to extinction.

5.5 PREY EXTINCTION

But what now happens if a predator variation appears that is so effective that it can drive the prey to extinction? We know that this can happen. For example, the invasive predator foxes and feral cats of Australia drove many prey species to extinction (Dickman, 1996). In the newt-snake war, snakes can ultimately beat the newts in TTX resistance. The data also shows that there are very few areas in the North American western coastal region with very low values of newt TTX (Brodie, 2010). In areas where the newt TTX is very low, it would not take much of a TTX-resistant snake invasion to nearly wipe out the newts. So what does our arms-race model predict when predator capabilities rise to the point where some prey have no defense?

So consider what happens when a predator variation appears that drives the prey countermeasure constant to zero. Remember Overfeeding Model 4.1. That case had a countermeasure constant of zero and the result was prey extinction.

Consider Figure 5.5a. The first timeframe starts with the ending of Predator Extinction Model 5.4. There are prey 4 and 7, predator 8, and countermeasure constants C48=4 and C78=8. Then, the advantage goes to a new predator variation in the next timeframe. Predator 9 appears with a variation that decreases C to C49=0 and C79=4. BARM is then exercised with the numbering and parameters shown in Figure 5.5b.

The results are shown in Figure 5.5c and the last timeframe of Figure 5.5a. Similarly to Predator Extinction Model 5.4, predator 9 has an extra resource to consume because of differences in C values. Predator 9 population grows.

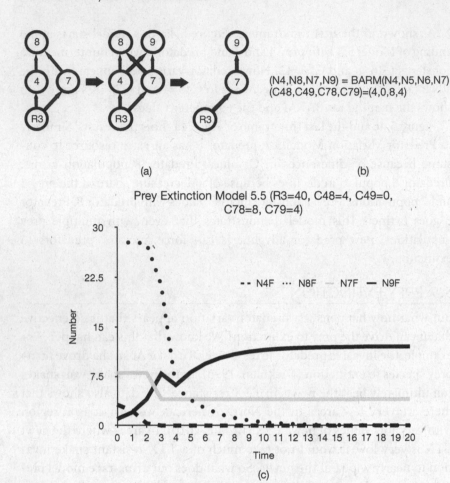

FIGURE 5.5   Prey Extinction Model 5.5. (a) The food-web timeframes show the beginning with the conclusion of Model 5.4, the appearance of an advantageous predator variation which zeros a prey C value, and the resulting extinctions. (b) The formulas show the BARM naming conventions and the parameter values. (c) The results show the extinctions of both prey and predator ancestors as well as the adaptation of the younger prey species to the new predator. The older prey species goes extinct because of no defense to the new predator. ˙

Predator 8 population declines because of the decreasing share of prey 4 and 7 populations. Prey 7 adapts to the C value with predator 9. Prey 4 goes extinct because of zero countermeasures. Predator 8 also goes extinct. This model demonstrates the extinction of both ancestor prey and predators when an advantageous predator variation appears, and a prey species has no defense.

5.6 RED QUEEN

So now we can finally ask whether there can really exist a Red Queen world where variations of predators and prey result in little or no apparent change. We've seen that predators are selected for improved capabilities against prey. We've also seen that prey without defenses can be selected against in favor of prey with defenses. Can a cyclic interaction occur with back-and-forth variations that realizes the Red Queen's hypothesis? This is the final question for this chapter. What can the models say about the Red

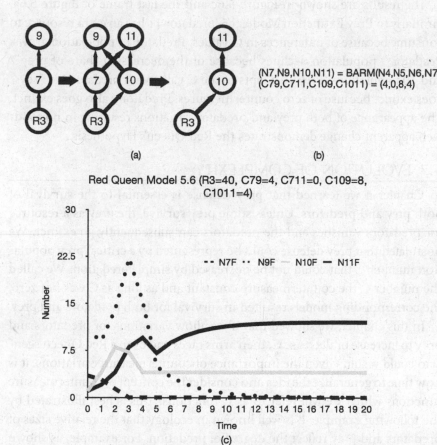

(a)

(N7,N9,N10,N11) = BARM(N4,N5,N6,N7)
(C79,C711,C109,C1011) = (4,0,8,4)

(b)

Red Queen Model 5.6 (R3=40, C79=4, C711=0, C109=8, C1011=4)

(c)

FIGURE 5.6   Red Queen Model 5.6. (a) The food-web timeframes show a beginning with the conclusion of Model 5.5, the appearance of both prey-advantage and predator-advantage variations, and the final result. (b) The formula show the BARM naming and parameters. (c) The results show the adaptation of the new prey to the new predator and the extinction of the old prey and predators. This demonstrates the Red Queen's Hypothesis.

Queen's Hypothesis? Is it necessary to adapt as quickly as possible just to stay in place?

Consider the food-web timeframes shown in Figure 5.6a. The first frame starts with the ending of Prey Extinction Model 5.5. There is prey 7, predator 9, and countermeasure constant C49=4. In the next frame, add a new prey variation 10 with an advantage that increases C to C109= 8. Also add a new predator variation 11 that decreases countermeasures to C711=0 and C1011=4. Exercise BARM with the values shown in Figure 5.6b.

The results are shown in Figure 5.6c and the last frame of Figure 5.6a. Similarly to Prey Extinction Model 5.5, predator 11 has an extra resource to consume because of differences in C values. Predator 11 population grows. Predator 9 population declines because of the decreasing share of prey 7 and 10 populations. Prey 10 adapts to the C value with predator 11. Prey 7 goes extinct because of zero countermeasures. Predator 9 also goes extinct. The appearance of both prey and predator variations resulting in no food-web apparent change demonstrates the Red Queen's Hypothesis.

5.7 EVOLUTION OF COMPLEXITY

In Chapter 4 we learned that prey defense is essential to the survival of both prey and predators. Unless some prey survive, the prey as a resource for predators vanishes and the predators can subsequently go extinct. We postulated that prey defense could be represented by a critical prey population number C that could not be decreased by simple predation. We called the number C the countermeasure constant and as long as C was not zero, the corresponding models resulted in survival for both predators and prey.

In this chapter, we showed that if we allow variations for predators and prey to increase or decrease C, then arms races and even a Red Queen scenario could result. Given the importance of countermeasure variations, it is now time to generalize the idea and consider the concept of countermeasure functions which we will refer to as C functions. This is best illustrated by the following example. It is well known in ecology that the relative sizes of predators and prey reflect the degree of predation. For example, as shown in Figure 5.7, animals in the Serengeti like elephants and hippos that are larger than a critical size are relatively free from predation.

However, animals below this size, such as impala and oribi, have significant predation (Sinclair et al., 2003). The critical weight is about 150 kg which interestingly is about the average weight of the adult females and males of the Serengeti apex predator, the lion.

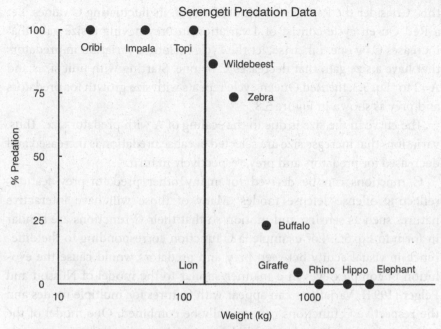

FIGURE 5.7   Serengeti Predation Data. Herbivore data from Sinclair, et al. (2010). Lion data is adult male and female average of East Africa data from Smuts, et al. (1980). Note that predation, as measured by the percent of animals that die from being killed by predators, drops from 100% for below the lion size to zero above it.

Obviously, the relative sizes of predators and prey matter. So consider the following C function

$$C = \max(\min(R, A(PREYSIZE-PREDATORSIZE) + B), 0) \quad (5.3)$$

where R is the prey carrying capacity, and A and B are constants. For the Serengeti predation data, B appears to be of the order of a fraction of R and A is of the order of R per a fraction of the predator size. The max and min functions keep the countermeasure function within physical bounds. This C function has the correct approximate behavior relative to the Serengeti predation data. The C function reflects significant predation for prey sizes much less than the predator size and little or no predation for prey sizes much larger than the predator size.

The key point of the countermeasure function Eq. 5.3 is that when it is inserted into this chapter's arms-race models, the appearance of predators and prey with incremental size variations results in the sizes of the predator and prey species growing automatically. A simple simulation can reflect this. Consider the Red Queen Model 5.6 with its fluctuating C values. Let a Red Queen cycle consist of a variation in prey having a size gain that increases C by one. Likewise, let the cycle include a variation in predators that have a size gain that decreases C by one. Starting with unit sizes and A=2 for Eq. 5.3, the Red Queen cycle repeats with size growth for predators and prey as shown in Figure 5.8.

The curve in the size is due to the scaling of A with predator size. Thus, variations that increase size are selected because predation is increased and decreased for predators and prey, respectively in turn.

C functions can be derived for many other predator-prey features reflecting offense-defense modes. Many of those will have interactive natures such as sensing and motion so that their C functions are similar in form to Eq. 5.3. For example, a C function corresponding to the difference in visual acuity between prey and predators would cause the evolution of complex eyes in a manner similar to the model of Nilsson and Pelger (1994). Variations can appear with features for multiple modes and the respective C functions can typically be combined. One model of the resultant combined C value would be the maximum of all the single-mode C functions. And as long as the ecology keeps changing and the ecology's resources continue to support, the features can grow like Figure 5.8 with the automatic production of complex evolution.

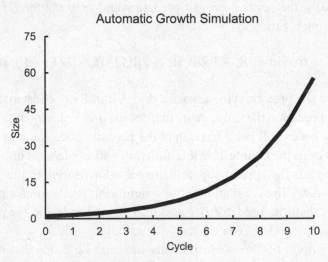

FIGURE 5.8 Simple simulation of automatic predator and prey size growth by adding C-formula Eq. 5.3 derived from Serengeti predation data to Red Queen Model 5.6. See text for discussion.

5.8 COMMENTS AND CONCLUSIONS

In general, arms races are neither symmetric nor reciprocal. The reason is simple. Predator variations are selected because, per the Selection Theorem, they have faster growth in limited prey resources. But for prey variations to be selected over older prey, the predators have to wipe the old prey out. This is not always easy to do. In addition, this constitutes an overfeeding condition that risks predator extinction if carried too far. Remember that prey defense was one of the key challenges in the previous chapter for the predators to overcome. Without prey defense, predators and prey can go extinct.

This situation shows that the Red Queen concept has some limitations. The Red Queen stage of the arms race is symmetric and reciprocal. But it must risk the potential for venturing into a situation with both predator and prey extinction. Arms races can proceed without the Red Queen stage, but they are not symmetric nor reciprocal.

One school of thought postulates that selection pressures are stronger on prey than on predators (Dawkins and Krebs, 1979). The argument goes like this. The hare runs faster than the lynx because the hare is running for its life while the lynx is only running for its lunch. This is referred to as the "Life-Dinner Principle." However, if the prey have built up sufficiently strong countermeasures, then contrary to the Life-Dinner Principle, selection pressure is stronger on predators than on prey. If a new prey variation appears with better countermeasures, there is relatively little selection. But if a new predator variation appears that decreases countermeasures, the difference acts like a new resource and the new predator population increases faster. Consequently, the old predators go extinct because of a decreasingly smaller share of resources consistent with the Selection Theorem. The data from Hague et al. (2020) and the models in this chapter are consistent with this.

Nevertheless, the models demonstrate that arms races are engines of evolutionary innovation and complexity. The models can also demonstrate the Red Queen's Hypothesis. Each change in the countermeasure constants reflects a new adaptation in predator or prey capability advantage. As long as the arms race continues, the capabilities will continue to grow. Advantageous adaptations could take numerous forms (mobility, morphology, sensory, etc.) as long as they are effective in altering countermeasures. This is the beginning of the complex ecologies phase of the models.

The stage is now set for an explosion of ecological diversity. What happens next is a chain reaction of events. An arms race starts between a predator and a prey. The variations are associated with one kind of mode that mutually affects countermeasures. The mode consists of the type of prey variation and corresponding predator response. The mode could consist of one of many different types of defense/offense tactics and responses such as mobility speed (each side running faster), size (each side getting larger), concealment/discovery, armor/anti-armor, and so on and on. The arms race will focus development along the selected mode and this will form distinguishing characteristics of the predator and prey species. Arms races will occur between herbivores and flora as well as between carnivores and fauna, whether carnivores or herbivores.

The point is that there are very many kinds of defense/offense modes (see e.g., Edmonds, 1974). Once a variation starts for something from one mode, the arms race ensues for that mode. For example, faster mobility from one side triggers even faster mobility from the other. Another variation, perhaps in another location, might pick a different mode and an arms race ensues for that one also. Soon, you have many different kinds of arms races, interaction occurs, and even more different arms races emerge. Even the transition from unicellular to multicellular life can occur from this (Herron et al. 2019). Complexity and novelty arise due to the different evolutionary trajectories resulting from the selection pressures of predator-prey offense/defense modes. An explosion of species diversity results.

This is what happened in the Cambrian explosion some 541M years ago. There is some discussion in the literature as to whether this process started in the Cambrian era or just before it (Wood, 2019). But the concept that the arms-race explosion occurred in biological evolution is gaining increasing acceptance.

CHAPTER 6

Trophic Cascades

NATURE ABHORS A VACUUM. If there is an ecological niche that can be filled, it will be filled. There is now an explosion in ecological diversity. Arms races of all kinds will proliferate in a never-ending struggle for survival. Variations in numerous forms will emerge in many ways such as for mobility, morphology, and sensory adaptations. Complexity and novelty will blossom where predators can become prey and prey can become predators. Hierarchies of food webs will appear. The Cambrian explosion is evidence of all this. And this is the way of the complex-ecologies phase.

At the top of the food web, there can be a species that has a very large effect on the rest of the ecology. Paine (1969) coined the term "keystone species" to describe apex predators at the top of the food web that are critical to the diversity of the ecosystem. The use of the term "keystone" is analogous to the Roman arch which would collapse if the keystone were removed. Similarly, removal of a keystone species would cause ecological collapse. Ecologies can be so interconnected that such removal can cause ecosystems to be categorized as "downgraded" (Estes et al., 2011).

Paine (1980) also coined the term "trophic cascades" to describe strong top-down effects when species were removed or reintroduced. Many trophic cascades have been observed in all kinds of natural habitats. Keystone species and trophic cascades are critical ideas towards understanding the

DOI: 10.1201/9781003391395-6

rules that determine the make-up of ecosystems (Carroll, 2016). So what can trophic cascades tell us about evolution and ecological change? And is it possible for downgraded ecologies to be upgraded? We will show that models can illuminate these concepts in this chapter and in Chapter 8.

6.1 TROPHIC CASCADE MODEL

In terms of the cuteness factor, sea otters have to be high on the list. They live in the ocean near the coast and feed on crustaceans and sometimes fish. They are one of the few mammals that use tools. They carry a rock in a pouch and use it to break loose food. They dive to the bottom, forage and dislodge food, and bring it to the surface. Then they lie on their backs on the surface, break open the food, and eat it. Although they have the thickest fur coats of any mammal they still need to consume about 1/4 to 1/3 of their body weight every day to compensate for heat loss. They forage starting in the early morning. Then the males float together in rafts for the afternoon grooming and siesta. The females float with their pups on their chests (Van Blaricom and Estes, 1988).

But sea otters have special ecological importance as a keystone species in a trophic cascade with sea urchins, ocean kelp, and orcas. Ocean kelp forests are common along the Pacific Northwest coast. To the coastal ecosystem, kelp forests are an important foundation for all the other species. Unfortunately, kelp forests are consumed when invaded by sea urchins. When sea otters invade, they consume the sea urchins and the kelp forests return. Without sea otters to protect them, kelp forests are vulnerable. Unfortunately, when orcas invade, they consume the sea otters, the sea urchins return, and the kelp forests recede again (Estes et al., 2010). We shall investigate this phenomenon using our previous models.

The Trophic Cascade Model (TCM) shown in Figure 6.1a is an extension of Predator Generalization Model 4.4. Predators are arranged in a trophic cascade. Numbering is restarted and the creatures have arbitrary complexity as described in Chapter 5. Number 1 is macroalgae (kelp) and number 2 is a predator (sea urchins). Number 3 is a predator (sea otters) and number 4 is an apex predator (orcas). All creatures are invasive species. TCM is modeled after the ecology of kelp, sea urchins, sea otters, and orcas of the Pacific Northwest coast.

The formulas for TCM are shown in Figure 6.1b.

TCM Details. Again, function g is used for formula compactness. First, kelp 1 consumes resource R1. Then consistent with

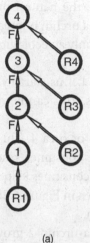

g(x,y)=max(0,x-y)
N1P=min(N1, R1)
N2P=min(N2, g(N1P, C1))
N2Q=N2P+min(N2-N2P,R2)
N3P=min(N3,g(N2Q,C2))
N3Q=N3P+min(N3-N3P,R3)
N4P=min(N4,g(N3Q,C3))
N1F=N1P-N2P
N2F=N2Q-N3P
N3F=N3Q-N4P
N4F=N4P+min(N4-N4P,R4)
N1=2 N1F
N2=2 N2F
N3=2 N3F
N4=2 N4F
Repeat above

(a)　　　　　　　　　　　　　(b)

FIGURE 6.1   Trophic Cascade Model (TCM). (a) The food web shows a trophic cascade that models the ecology of the Pacific Northwest Coast. 1=kelp, 2=sea urchins, 3=sea otters, and 4=orca. Resources R2, R3, and R4 provide resource generalization. (b) The formulas are based on those for Model 4.4.

countermeasure constant C1, sea-urchin 2 consumes kelp 1, unfed sea-urchin 2 feeds on resource R2, and the result is totaled. Sea-otter 3 consumes sea-urchin 2 consistent with countermeasure constant C2. Unfed sea-otter 3 feeds on resource R3 and the result is totaled. Orca 4 feeds on sea-otter 3 consistent with countermeasure constant C3. Unfed orca 4 feeds on resource R4. All the consumption numbers are totaled, and replication takes place.

6.2 LOW-TIER PREDATION

The collapse of a forest is an ecological catastrophe. When the American chestnut forests died from disease the black bears that depended on the chestnuts for hibernation food also disappeared from those forests. Other species suffered too (Stupka, 1960; Diamond et al., 2000). The same ecological catastrophe is now happening in the world's oceans—the kelp forests are dying. As a result, numerous fish and invertebrate populations are collapsing. A 2016 study (Krumhansl et al., 2016) noted that

Kelp forests are increasingly threatened by a variety of human impacts, including climate change, overfishing, and direct harvest.

For example, in several parts of the world, the natural predators of sea urchins have been reduced. The result is a sea urchin population explosion that wipes out entire kelp forests and leaves only an ecological wasteland of urchin barrens (Johnson et al., 2011).

What can our models tell us about this? Let us start by looking at the invasion of the kelp beds by the sea urchins. This can be modeled by the food web shown in Figure 6.2a.

This model is TCM with no sea otters 3 or orca 4. This model is also Predator Generalization Model 4.4. Kelp 1 consumes resource R1. Sea-urchin 2 is an invasive species. Sea-urchin 2 consumes kelp 1 first and then resource R2. The formulas for TCM are shown in Figure 6.2b and again use the previously employed naming convention.

The results are shown in Figure 6.2c. Sea-urchin 2 grows due to consuming kelp 1. Kelp 1 diminishes because it is consumed. Kelp 1 is eventually almost totally suppressed. This model demonstrates that predation can cause suppression.

6.3 MID-TIER PREDATION

In the early eighteenth century, the sea otter population of about 200,000 ranged along the North Pacific coasts from Japan to Baja California. Then in 1779, the expedition of James Cook stopped in Canton China and auctioned off some sea otter pelts that they had obtained from Vancouver Island. The bidding went so high that the word quickly spread of the high value of these pelts. This triggered a stampede of sea otter hunting that lasted until it was banned by international law in the early twentieth century. By the time it was over, the sea otter was all but extinct with a population of only about 1000. During this time, the sea urchins thrived and the kelp forests died. Then the sea otters started to recover, first in the Aleutian Islands. Starting in 1949, efforts were made to transplant sea otters to other locations along the North American west coast (Kenyon, 1969). Some efforts of re-introduction were more successful than others. But where they were successful an interesting thing happened—the sea otters devoured the sea urchins and the kelp forests returned (Estes et al., 2010).

Now let's see what happens in the models when the sea otters show up. Consider the food web shown in Figure 6.3a. This is TCM without orca 4. Invasive species predator 3 (sea otters) is added to the ending of Low-Tier Model 6.2. Sea otter 3 consumes sea urchin 2 first. Then sea otter 3 consumes resource R3. The sea urchin 2 countermeasure constant is set at C2=0.4 as shown in Figure 6.3b.

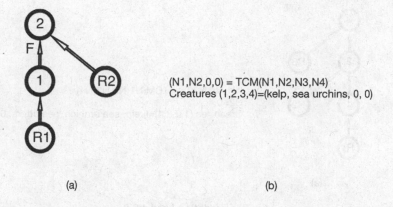

$(N1,N2,0,0) = TCM(N1,N2,N3,N4)$
Creatures $(1,2,3,4)=$ (kelp, sea urchins, 0, 0)

(a)   (b)

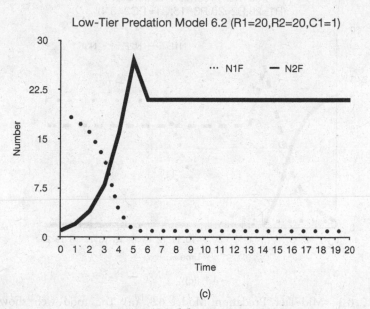

(c)

FIGURE 6.2   Low-Tier Predation Model 6.2. (a) The food web is TCM without sea otters 3 and orca 4. It is also Model 4.4. (b). The formulas reflect the naming conventions. (c) The results show that the invasive species 2 (sea urchins) suppresses the native species 1 (kelp). Please note again that the selection of parameters and corresponding number scales are for concept demonstration purposes only.

The results are shown in Figure 6.3c. Sea otter 3 population rises. Sea urchin 2 population is suppressed. Kelp 1 population is now much less suppressed than before and rises. This model demonstrates sea otter 3 as a keystone species. This model also demonstrates a trophic cascade with strong top-down effects with the presence or absence of sea otter 3.

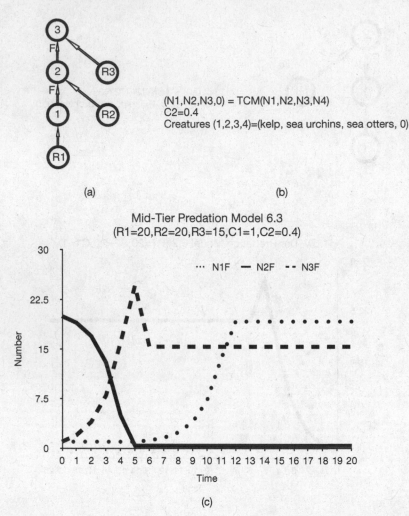

(a)

(b)

$(N1,N2,N3,0) = TCM(N1,N2,N3,N4)$
$C2=0.4$
Creatures (1,2,3,4)=(kelp, sea urchins, sea otters, 0)

(c)

FIGURE 6.3   Mid-Tier Predation Model 6.3. (a) The food web shows the appearance of invasive species predator 3 (sea otters) to Model 6.2. (b) The formulas show TCM naming and parameters. (c) The results show the rise of sea otter 3, the decline of sea urchins 2 and the rise of kelp 1 that had previously been suppressed. This model demonstrates both keystone species and trophic cascades.

6.4 APEX PREDATOR

Just as things were looking up for the sea otters, in 1990 the roof fell in (Estes et al., 2009). Orca (killer whales) had relied on great whales as one of their key food resources. With the commercial destruction of the great whales by 1975 in the North Pacific, the orca had to shift foraging strategies to smaller prey. The problem was that orca food requirements were large and shifting to smaller mammals meant that they had to eat a lot of them. What happened next was the destruction by orca of a succession of ocean mammal populations. First the harbor seals, then the fur seals, then the sea lions, and finally the sea otters. About 10,000 sea otters vanished annually in the Aleutian Islands alone. As small as a group of five orca could have caused the sea otter such a near extinction in the 1990s (Williams et al., 2004). As expected, the loss of the sea otters brought back the sea urchin barrens and the loss of the kelp forests.

The orcas 4 appear as shown in Figure 6.4a. This model adds apex predator 4 (orca) to the ending of Mid-Tier Model 6.3. This model is also the full TCM. Apex predator 4 (orca) is an invasive species. Orca 4 consumes sea otters 3 first. Then orca 4 consumes resource R4. The sea otter 3 countermeasure constant is set at C4=0.1 as shown in Figure 6.4b.

The results are shown in Figure 6.4c. Orca 4 population rises. Sea otter 3 population declines. The previously suppressed sea urchin 2 population now rises again because its suppression has been in turn suppressed. In consequence, kelp 1 population is suppressed again. This model demonstrates apex predator 4 (orca) as a keystone species with a trophic cascade that strongly affects the levels below it.

> TCM Model Details. Please note that the choice of countermeasure constants is not accidental. Predators suppress prey to their C values unless a higher predator drives the lower predator to a C value that permits the prey to bloom. This happens when the lower predator population does not consume all the prey above the prey C value. In our case after replication, when $(N1-C1)>N2$, $N1=2xC1$, and $N2=2xC2$, then if $C1>2xC2$, number 1 will bloom after suppression of number 2 by number 3. The same logic applies to numbers 2 and 3 with blooming for $C2>2xC3$ after suppression by number 4. These prey blooming thresholds due to suppression of predators by higher predators can be generalized for arbitrary replication ratios with $(K1-1)$ $C1>K2xC2$ for replication ratios K1 and K2 for numbers 1 and 2 respectively.

6.5 LOW-TIER COMPETITION

One of the great delights of visiting the seashore is walking along the beach and seeing the tidal pools. They frequently contain whole zoos of animals from starfish to clams and snails. Some places are relatively pristine and

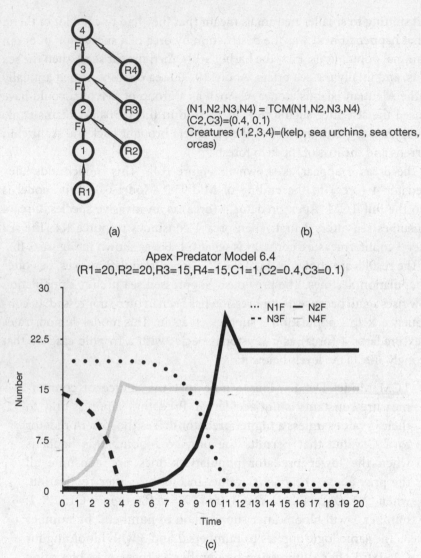

(N1,N2,N3,N4) = TCM(N1,N2,N3,N4)
(C2,C3)=(0.4, 0.1)
Creatures (1,2,3,4)=(kelp, sea urchins, sea otters, orcas)

(a)   (b)

(c)

FIGURE 6.4   Apex Predator Model 6.4. (a) The food web show the apex predator 4 (orcas) invading the food web of Model 6.3. (b) The formulas reflect the TCM naming and parameters. (c) The results show the bloom of orca 4, the crash of sea otter 3, the bloom again of sea urchin 2, and the crash again of kelp 1. This model demonstrates apex predators as keystone species with strongly affected trophic cascades.

have many such pools, all lying near each other and filled with beautiful sea creatures. One such place is Makkaw Bay along the ocean shore of Washington State. Bob Paine found the location beautiful too. But he also realized that it offered a chance to make a change in one pool and leave another alone to see what the difference in outcome might be. So he decided to try "kick-it-and-see" experiments by removing one species from one pool at a time. There were lots of pools, so he could do this in parallel with each of several pools missing one different species and another pool left alone. He watched for years, but in almost all the pools, nothing happened.

But when he removed the starfish, everything changed. The mussels took over completely and the rest vanished. The reason why was that the starfish was an apex predator and a keystone species that ate and controlled the mussels. When the starfish were removed, the mussels crowded everyone else out, a trophic cascade followed, and the ecosystem was downgraded (Paine, 1966, 1969, 1974, and 1980). This work was the original impetus for the realization of the importance of keystone species and trophic cascades (Carroll, 2016).

The trophic cascade of the intertidal habitat is distinct from that of the sea otters because the lowest tier is governed by competition rather than predation. The mussels are competing against the rest of the animal zoo instead of the sea urchins eating the kelp. So we will modify our TCM Model to create an alternate TCM Model that reflects this low-tier competition.

So what does happen if the low-tier creatures compete for resources rather than be predators and prey? Consider Figure 6.5a which shows a low-tier resource-competition as an alternative to just a predator-prey cascade.

The alternate TCM model is a combination of Selection Model 1.3 and Predator Model 4.4. Zoo creatures 1 and mussel 2 compete for resource R1. Mussel 2 consumes resource R1 first and then resource R2. Starfish 3 is a predator. The term "zoo" represents the flora and fauna (over 25 species) without the mussels and starfish. The formula modifications are shown in Figure 6.5b.

When starfish 3 is removed at time=5, the results are shown in Figure 6.5c. The previously suppressed mussel 2 population now blooms because its suppression has been removed. In consequence, the zoo 1 population is suppressed. This model demonstrates starfish predator 3 as a keystone species with a trophic cascade that strongly affects levels below it that include resource competition.

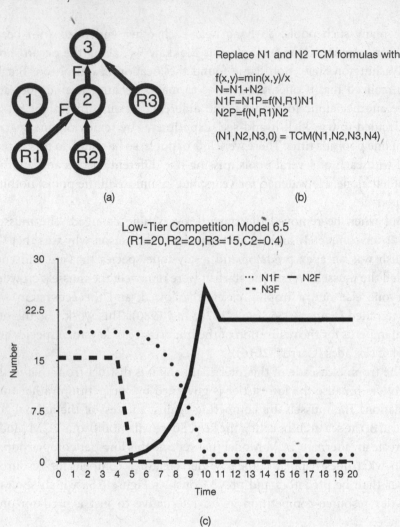

Replace N1 and N2 TCM formulas with

$f(x,y) = \min(x,y)/x$
$N = N1 + N2$
$N1F = N1P = f(N,R1)N1$
$N2P = f(N,R1)N2$

$(N1,N2,N3,0) = TCM(N1,N2,N3,N4)$

(a)   (b)

Low-Tier Competition Model 6.5
(R1=20,R2=20,R3=15,C2=0.4)

··· N1F   — N2F
-- N3F

(c)

FIGURE 6.5   Low-Tier Competition Model 6.5. (a) The food web shows a trophic cascade that models the intertidal habitat in Makkah Bay, Washington and the experiment of Paine (1966). 1=zoo, 2=mussels, and 3=starfish. Resources R2, and R3 provide resource generalization. (b) The formulas reflect the modifications to TCM whereby low-tier predation is replaced with competition for zoo 1 and mussel 2. (c) Starfish 3 is removed at time=5. The results show the bloom of mussel 2 and the crash of zoo 1. This model demonstrates apex predators as keystone species with strongly affected trophic cascades that include low-tier competition for resources.

6.6 COMMENTS AND CONCLUSIONS

The Trophic Cascade Models (TCM and alternate) demonstrate keystone species and trophic cascades for both predator-prey and resource-competing lower levels. Many biological ecologies, from starfish to wolves, behave as keystone species and trophic cascades. Such behavior shows that while biological robustness may exist for many species (e.g., bacteria), it is not universal for ecologies. Evolution is about adaptation not necessarily robustness. Robustness would be more likely seen in older species with large exposures to very many different environments.

Extinctions and near-extinctions are signature events of evolutionary selection. Whatever the ecological gene pool was beforehand, it is certainly different afterwards. Whereas this is obvious for extinctions, it is also true for near-extinctions and is known in evolutionary theory as the founder effect (Mayr, 1942). If forced through a bottleneck, remnant populations will reflect a loss of gene diversity for a very long time. This is exactly what happened to the sea otter gene pool (Larson et al., 2002).

Extinctions and near-extinctions can happen relatively quickly, especially with predators involved. The disappearance of several species in Australia due to the introduction of foxes and feral cats happened in a small number of generations (Dickman, 1996). The Canadian lynx and snowshoe hares repeatedly brush with near extinction about every ten years (Krebs et al., 2001). The reason why evolution sometimes takes time is the need for a new gene to appear. This was certainly reflected in Lenski's long-term *E. coli* experiment (Lenski, 2017). But if the gene needed to survive is already in the gene pool, selection can occur relatively quickly as shown by the Selection Speed Theorem of Chapter 1 and the arms races models of Chapter 5.

The extinction and near-extinctions of trophic cascades are also evolutionary selection. And the speed of their progression in time is not inconsistent with evolutionary processes. Therefore, trophic cascades can be viewed not only as ecological change but also as evolution in action.

Although trophic downgrading appears to reduce complexity, trophic upgrading offers the potential for increased complexity such as with the return of the kelp forests. It is no accident that the conceptual models derived from the previous chapters also show conceptual relevance to trophic cascades. Thus, our models are providing illumination for the rules of evolution and ecology. Ecological change is the engine of evolutionary novelty and predators play a central role.

CHAPTER 7

Parasites and Pathogens

NOWHERE IS EVOLUTION SO manifest as in the war between hosts and infectious diseases from parasites and pathogens. Like predators, infectious diseases can cause extinctions. But because they can mutate and replicate much faster than hosts, their effects on evolution can be much stronger. A mammalian predator can cause extinction but must act over years to build up a population to do so. But an infectious disease can bloom at such speed that it can cause extinction within one host generation. Such species extinctions by infectious diseases have been implicated in the disappearance of many species (Smith et al., 2006).

Similarly to defense against predators, infectious diseases have forced hosts to develop multiple countermeasures such as resistant genes, immune systems, and enhanced fertility. For example, the European wild rabbit can breed in three or four months, produce a litter of about five babies in about a month, and breed again immediately. As long as conditions allow, they can keep up this pace and give birth to scores of baby rabbits per year. Unfortunately in 1859, some Australians forgot about this and released 12 pairs of European wild rabbits for sport hunting. With a continent of available rabbit food and relatively little predation, the consequence was an agricultural and ecological catastrophe. By releasing the rabbits without the

DOI: 10.1201/9781003391395-7

predators that controlled their native population, the Australians created a trophic cascade with the consequence of a downgraded ecosystem.

Within ten years, two million could be killed without having a noticeable effect on the rabbit population. By 1950, the Australian rabbit population was estimated at 600 million. At this time, the human population of Australia was only about 8 million, so the rabbits outnumbered the humans by about 75 to 1. At this level, an Australian human family of four with a typical suburban plot of a quarter of an acre of land could have had a rabbit every two yards.

The Australian government was desperate to do something. Conventional control methods (hunting, trapping, fencing, poisoning, etc.) had all failed miserably. So the government finally decided to deploy the infectious disease MYXV, a virus, against the rabbits. The results were one of the greatest scientific examples of evolution and coevolution (Kerr et al., 2015).

7.1 VIRULENCE

One of the problems with the MYXV disease selected by the Australians was its extreme virulence (host lethality). If virulence is too low, a disease may have too little chance to multiply and be transmitted. But if virulence is too high, the host will die before it has a chance to spread. This was the problem with the original strain of MYXV known as SLS (Standard Laboratory Strain). MYXV was first discovered at the Pasteur Institute in Montevideo, Uruguay in 1896 when their European laboratory rabbits all died. In 1919, the Brazilian scientist Dr. H. Aragao recommended to the Australian government the use of MYXV as a solution to their rabbit problem. This recommendation was repeated by the Australian physician-scientist Dr. Jean Macnamara in 1934. Australian studies commenced but although it killed all captive rabbits, it could not be spread into the wild because of its extreme virulence. Studies were postponed during World War II but re-started in the Australian winter (May) of 1950. Again the captive rabbits died and the disease did not spread to the wild. But in the Australian spring (December) of 1950, a bloom of mosquitoes transmitted MYXV to the wild rabbit population and the result was nearly 100% lethality (Fenner, 1983).

Parasites and pathogens mutate and multiply very much faster than hosts. Over the host replication timescale, which is a generation, infectious diseases act like a complex of diseases that adds an additional effective mortality rate. This complex is the sum of all infectious diseases for each generation. Thus, parasites/pathogens and hosts engage in arms races. In this chapter, we will explore this interaction between infectious diseases and hosts.

In order to establish some fundamentals, we begin our exploration of infectious-disease effects with a simple model. Start with replicator Model 1.1. As shown in Figure 7.1a, add an additional effective mortality rate $M1$ to represent the effect of an infectious-disease complex. The mortality rate can vary in time due to the appearance of changes in the complex. Hosts can also create variations with improved mortality rates to reflect improvements in resistance (better immune systems, etc.). The formulas are shown in Figure 7.1b. As usual, K is the replication ratio.

What happens if the pathogen mortality $M1$ is too large? If you examine the formulas of Figure 7.1b, the host population decreases with time if

$$(1-M1)K < 1 \text{ or equivalently} \qquad (7.1)$$

$$M1 > 1 - 1/K \qquad (7.2)$$

Thus, too much pathogen mortality causes host species extinction. For an infectious disease that is species specific, this can also lead to parasite/pathogen extinction analogous to predator overfeeding as discussed in Chapter 4. So there is a limit to virulence. Unfortunately, too many infectious diseases can jump species and avoid extinction in a manner analogous to predator generalization. In that case, only the host goes extinct. Please also note the host sensitivity to infectious-disease-induced mortality for low-fertility (i.e., low K) species.

Figure 7.1 presents a computational example of infectious-disease-induced extinction. We set $K=2$ and the initial mortality rate to be $M1=0$. At time=3, an infectious disease appears with a large mortality rate of $M1=0.6$. The results are shown in Figure 7.1c. The host 1 population is correspondingly decreased by the new mortality rate and proceeds to extinction.

7.2 ATTENUATION

Viruses mutate and thereby create new strains. MYXV is a DNA virus and so mutates less than an RNA virus, but it still mutates and makes new strains. This sets up a competition among the strains and the ones with the greater growth dominate while the ones with less growth are suppressed.

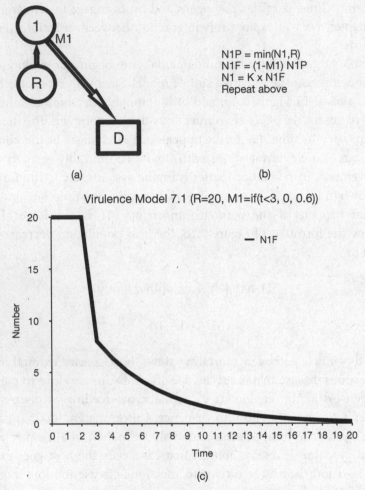

N1P = min(N1,R)
N1F = (1-M1) N1P
N1 = K x N1F
Repeat above

(a)   (b)

Virulence Model 7.1 (R=20, M1=if(t<3, 0, 0.6))

(c)

FIGURE 7.1   Virulence Model 7.1. (a) The food web shows an extra mortality M1 from infectious disease. (b) The formulas show the modeling of the extra mortality M1. (c) At t=3, there appears an infectious-disease complex with a large mortality M1=0.6. The result is extinction for host 1.

Virus growth is determined by the speed of replication and transmission. Faster replication will make a virus more virulent but make for slower transmission by killing the host. But slower replication can increase transmission from host to host by allowing the host to live longer and give more time for transmission. This phenomenon of virus mutation and virulence reduction is known as attenuation and is often observed in epidemics (Anderson and May, 1982). In the first years of the MYXV epidemic, attenuation is exactly what happened. The extremely virulent SLS strain was suppressed and less virulent strains became prevalent (Fenner, 1983; Ridley, 2004, pp. 625–626; Peng et al., 2016).

Consider the food web in Figure 7.2a, and the formulas in Figure 7.2b, which are identical respectively to those in Figure 7.1a and Figure 7.1b.

The model computation proceeds as follows. Again K = 2. Repeat Model 7.1 with M1=0 at t=0 and M1=0.6 at t=3. Then set at t= 6, the infectious disease changes to a mortality rate of M1=0.35. The results are shown in

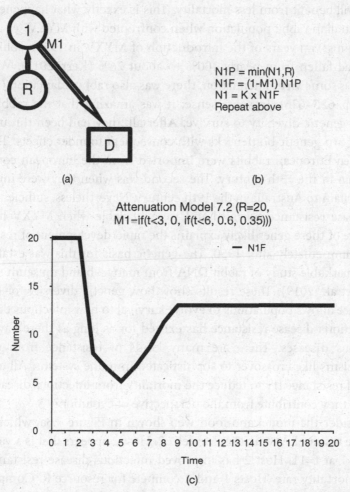

(a)

N1P = min(N1,R)
N1F = (1-M1) N1P
N1 = K x N1F
Repeat above

(b)

Attenuation Model 7.2 (R=20,
M1=if(t<3, 0, if(t<6, 0.6, 0.35)))

— N1F

(c)

FIGURE 7.2 Attenuation Model 7.2. (a) The food web is the same as Virulence Model 7.1 (b) The formulas are the same as Virulence Model 7.1. (c) The results repeat Model 7.1 with M1=0 at t=0 and M1=0.6 at t=3. But then attenuation occurs with M1=0.35 at t=6. The host 1 population correspondingly recovers because the infectious disease attenuated.

Figure 7.2c and the host 1 population correspondingly survives with the new attenuated mortality rate.

7.3 RESISTANCE

Genetic diversity is a central bulwark against infectious diseases (King and Lively, 2012). As long as a species has some genes that can provide resistance to an epidemic, that species will survive. The ones that don't have resistant genes can suffer significant mortality, but the ones with some resistant genes will benefit from less mortality. This is exactly what happened with the Australian rabbit population when confronted with MYXV.

Within seven years of the introduction of MYXV in 1950, rabbit mortality had fallen from nearly 100% to about 26% (Kerr, 2012). Whereas there was some virus attenuation, there was also rabbit adaptation (Ridley, 2004, pp. 625–626). In some sense, it was amazing that the rabbits had enough genetic diversity to survive. After all, they had been through not one but two genetic bottlenecks with consequent founder effects. The first was when European rabbits were imported from the European continent to Britain in the 13th century. The second was when they were imported from Britain to Australia in the 19th century. Nevertheless, sufficient genes for disease resistance survived and were available when MYXV hit. The presence of these genes likely explains the rapid development of resistance almost immediately after 1950. The genetic basis for this was established by a remarkable study of rabbit DNA from many lab and museum sources (Alves et al., 2019). These results show how genetic diversity for disease resistance allows populations to evolve survival to new infectious diseases.

Infectious-disease resistance has existed for as long as there have been infectious diseases. There are many kinds of resistance from genetic mechanisms like crossover to sophisticated immune systems. All of these kinds of resistance try to reduce the mortality from infectious diseases. So, how do they contribute from the perspective of evolution?

Consider the model and food web shown in Figure 7.3a, which starts with the result of Attenuation Model 7.2. Host 2 appears that is a variation of host 1 at t=11. Host 2 has improved infectious-disease resistance and lower mortality rate. Hosts 1 and 2 compete for resource R. Competition is modeled in the usual way via Selection Model 1.3. The formulas in Figure 7.3b reflect this.

The results are shown in Figure 7.3c. The initial progression is the same as Model 7.2. But at t=11, the new resistant host appears and its population

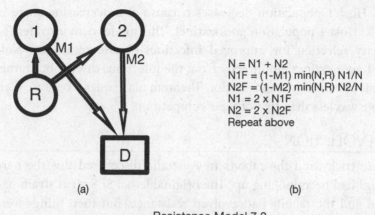

(a)

$N = N1 + N2$
$N1F = (1-M1) \min(N,R) \, N1/N$
$N2F = (1-M2) \min(N,R) \, N2/N$
$N1 = 2 \times N1F$
$N2 = 2 \times N2F$
Repeat above

(b)

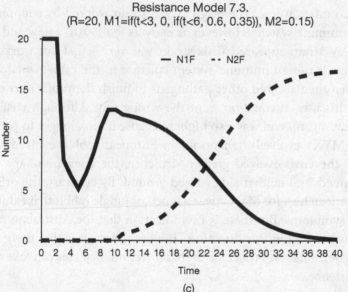

Resistance Model 7.3.
(R=20, M1=if(t<3, 0, if(t<6, 0.6, 0.35)), M2=0.15)

(c)

FIGURE 7.3 Resistance Model 7.3. (a) The food web shows Model 7.2 with the appearance of a variation host 2 that shares resource R and has improved pathogen mortality M2. (b) The formulas show the usual resource sharing but with different pathogen mortality. (c) The results show the same initial progression as Model 7.2 but the emergence at t=11 of host 2 that has better disease resistance. This model demonstrates the selection of hosts with better disease resistance. Note that the extinction of host 1 is not directly due to the disease but rather to classic selection from a decreasing share of limited resources.

increases. Host 1 population decreases because of a decreasing share of resource R. Host 1 population goes extinct. This model demonstrates the evolutionary selection for improved infectious-disease resistance. Note that host 1 goes extinct not directly from the infectious disease but rather from classic selection via the Selection Theorem of Chapter 1 because their growth rate was less than that of their competition.

7.4 COEVOLUTION

Viruses are tricky and the rabbits in Australia discovered this the hard way. Things had been looking up. The original lethal SLS virus strain had attenuated and the rabbits had evolved resistance. But then things went from bad to much worse. The original SLS strain worked by suppressing the rabbit immune system. However as early as the 1970s, a new and very nasty MYXV strain appeared. This strain was very lethal but worked in a tricky way. It caused immune-system collapse in the rabbits analogous to AIDS that then allowed other pathogens to finish them off (Kerr et al., 2017). In this way, attenuation went the wrong way. Although virulence was higher, transmission was also higher because it took longer to perish.

At first, MYXV evolved attenuation, the European rabbits evolved resistance, then the virus evolved greater virulence, the coevolutionary arms race continued, and neither side gained ground. By contrast, the original South American host for MYXV was a tapeti or jungle rabbit that exhibited only mild symptoms. But there is no indication that the Australian rabbit epidemic is heading in this direction. The broad lesson is that there are a variety of coevolutionary trajectories that infectious diseases can take other than coexistence.

Parasites/pathogens and hosts have been engaged in a never-ending arms race since the dawn of life itself. Just like arms races between predators and prey as described in Chapter 5, parasites/pathogens and hosts constantly try to gain an advantage over the other in the struggle for survival. The result has been ever more complex infectious diseases and disease resistance. So, how does this happen from an evolutionary perspective?

Consider the model and food web shown in Figure 7.4a. Start with the initial progression of Resistance Model 7.3. A new pathogen complex appears at t= 35 with higher mortality. The formulas in Figure 7.4b mirror that of Model 7.3.

The results are shown in Figure 7.4c. The new infectious-disease complex appears at t=35 that raises mortality of host 2 from 0.15 to 0.45. Host

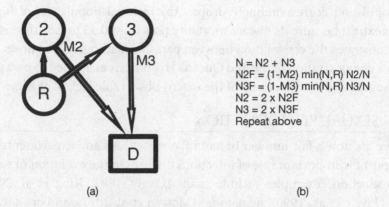

N = N2 + N3
N2F = (1-M2) min(N,R) N2/N
N3F = (1-M3) min(N,R) N3/N
N2 = 2 x N2F
N3 = 2 x N3F
Repeat above

(a)   (b)

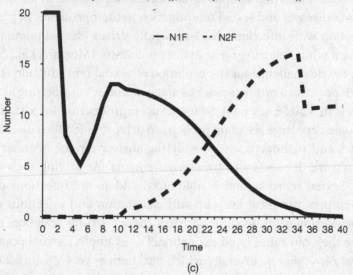

(c)

FIGURE 7.4   Coevolution Model 7.4. (a) The food web starts with Model 7.3 and adds a new disease strain with greater virulence. (b) The formulas mirror that of Model 7.3. (c) The new disease strain with M1=0.6 and M2=0.45 appears at t=35. The result shows the initial progression from Model 7.3 and the consequence of the new virulent strain appearance at t=35. This model demonstrates the effects of coevolution between infectious diseases and hosts. The figure also shows the complete sequence for the beginnings of virulence (t=3), attenuation (t=6), resistance (t=11), and virulence coevolution (t=35).

2 population correspondingly drops. Any residual population of host 1 goes extinct because its disease mortality rises from 0.35 to 0.6. This model demonstrates the coevolution between parasites/pathogens and hosts. It is also a demonstration of the Red Queen's Hypothesis as applied to parasites/pathogens and hosts because of the macro-observable lack of change.

7.5 SEXUAL REPRODUCTION

There are now a fair number of natural observations and experiments that support the important role of infectious diseases in the evolution of sexual reproduction. Examples include snails (Lively, 1987; King et al. 2009), fish (Lively et al., 1990), nematodes (Morran et al. 2011), and crustaceans (Ganz and Ebert, 2010). In all of these studies, the competition between asexual and sexual populations favors asexual replication in the absence of infectious diseases and sexual reproduction in their presence. In particular, coevolution with infectious diseases rapidly drives asexual populations to extinction while selecting for sexual reproduction (Morran et al., 2011).

The classic challenge for the evolution of sexual reproduction is that an asexual species can produce twice as many reproducing offspring as a sexual species with a 50:50 sex ratio. So for sexual reproduction to exist, there has to be some very large selection pressure to drive it. Infectious diseases from parasites and pathogens can provide the answer for the necessary selection pressure. It meets all of the requirements for driving the evolution of sex. Sexual reproduction is ubiquitous and so are infectious diseases. Selection pressure must keep up with adaptation and infectious diseases are coevolutionary. Infectious diseases induce strong selection pressure because they can cause rapid extinction. For example, asexual populations have very low genetic diversity which put them at very great risk of rapid extinction from infectious diseases. In addition, the rapid creation of genetic diversity by sexual reproduction is clearly a powerful weapon against infectious diseases. Thus, infectious diseases have the potential for the selection pressure needed to drive the evolution of sexual reproduction.

However, this begs another question. Evolution works because it takes relatively small steps and proceeds gradually. Large steps are usually fatal because too many things have to work at once. So the question is: what are the gradual steps going from an originally asexual replication to sexual reproduction?

Consider the following model. As illustrated in Resistance Model 7.3, selection in infectious-disease environments is set by the effective replication ratio KE=(1-M)K where M is the mortality and K is the replication ratio. With KE1 and KE2 the effective replication ratios for hosts 1 and 2 respectively, host 2 is selected over host 1 if

$$KE2 >= 1 \text{ and} \tag{7.3}$$

$$KE2 > KE1 \tag{7.4}$$

In keeping with the Selection Theorem, if the population of host 2 grows faster than the competition and does not go extinct, it will be selected. Sexual reproduction decreases the replication ratio because only females can reproduce so the replication ratio K is replaced by the reproduction ratio KP where

$$KP = 1 + (K-1)/(1+SR) \text{ and} \tag{7.5}$$

$$SR = \text{Sex Ratio} = \text{Males / Females} \tag{7.6}$$

Now apply this model to competition between asexual and sexual hosts with different pathogen mortalities. Figure 7.5 shows 5 regions for values of mortalities MS (sexual) and MA (asexual) and K=2. Note that the case of polygamy with 0<SR<1 is also considered. The regions and their consequences are described as follows:

A. Nobody wins for MA>1/2 and MS>1/2,

B. Asexual wins for MA<1/2 and MA<MS,

C. 50:50 sexual wins for MS<1/3 and MS<1-(1-MA)4/3,

D. Depending on SR, asexual or polygamy (0<SR<1) can win in region for MA<1/2, MA>MS and MS>1-(1-MA)4/3, and

E. Depending on SR, only polygamy can win for MA>1/2 and 1/3<MS<1/2.

Consistent with expectations, there is a large gap between asexual replication and 50:50 sexual reproduction that would take considerable selection pressure to cross with high asexual mortality and much lower sexual mortality. In addition, Figure 7.6 shows the detailed competition boundaries between asexual and polygamous hosts for various sex ratios SR. Low SR polygamy can be a small initial step from asexual reproduction.

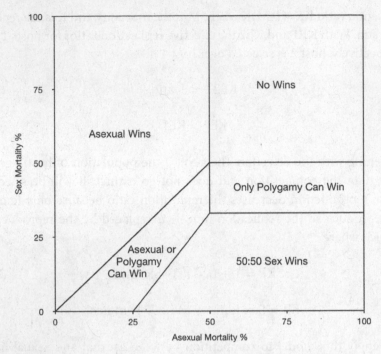

FIGURE 7.5 Infectious-disease-mortality tradeoff for competition between asexual replication and sexual reproduction with the replication ratio K=2. Regions are determined by differential effective replication ratios. For example, nobody wins and all go extinct if both mortalities are more than 50%. Note the large gap between asexual and 50:50 sexual. See text for discussion.

So how can evolution proceed to select for sexual reproduction? The key is the presence or absence of disease-resistant genes and their determination of disease-induced mortality. When a mortality-inducing infectious disease appears, there can be selection for sexual reproduction over asexual replication. As long as the number of surviving sexual female offspring is more than the corresponding number of surviving asexual offspring, the sexual population will grow faster, and, per the Selection Theorem, the asexual population will go extinct.

There are two scenarios in which this can occur. The first is when the sexual population has disease-resistant genes and the asexual one does not. This can happen because of the low genetic diversity of asexual populations and the potential for larger genetic diversity for sexual populations. In particular, low SR populations have effective replication ratios close to that of the asexual population. Consequently, the amount of shift in mortality

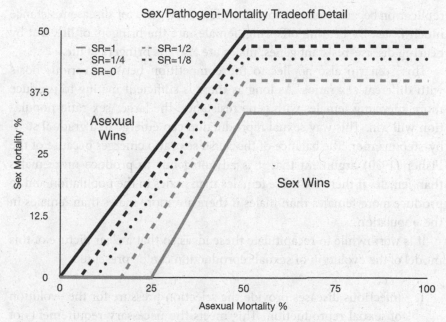

FIGURE 7.6 Sex-ratio detail of infectious-disease-mortality tradeoff for competition between asexual replication and sexual reproduction. Regions are determined by differential effective replication ratios for different sex ratios SR= NumberOfMales/NumberOfFemales. Note how close low SR polygamy is to asexual replication. See text for discussion.

necessary for selection of sexual reproduction can be small. For example, the effective replication ratio KE for the sexual population will increase if

$$\Delta M < - F \, \Delta SR \qquad (7.7)$$

where ΔM is the shift in disease-induced mortality and

$$F = (1-M)(KP-1)/((1+SR)KP) \qquad (7.8)$$

But the second scenario truly highlights the importance of sex. This is when both asexual and sexual populations have the same relatively rare gene for resistance. This is the scenario where the males are the drivers. If the males mate enough times so that the number of surviving female offspring with gene resistance is more than the corresponding number of asexual offspring, the sexual population growth rate will be larger and the sexual population will win. Sexual reproduction is selected over asexual replication because sex can spread a life-saving gene for disease resistance much faster. By bearing offspring, females are the bringers of life. But by bearing disease-resistant genes, males are also the bringers of life.

This scenario also applies to the competition between various hosts with different sex ratios. As long as there is sufficient mating to produce more surviving females with gene resistance, the larger sex-ratio population will win. This way sexual reproduction can emerge in a gradual step-by-step manner. The balance of the 50:50 sex ratio emerges because of the Fisher (1930) argument that it is advantageous to produce more males than females if there are more females than males in the population and to produce more females than males if there are more males than females in the population.

It is worthwhile to recapitulate these ideas so that a clear picture of this model of the evolution of sexual reproduction can be presented.

1. Infectious diseases provide the selection pressure for the evolution of sexual reproduction. This meets the necessary requirements of strength, frequency, and adaptability. Infectious diseases are a bane to asexual species with low genetic diversity and can coexist with genetically diverse sexual species. Numerous empirical studies support the concept of infectious diseases as the driver for the evolution of sexual reproduction.

2. Infectious diseases provided a long sequence of strong selection events whereby populations with disease-resistant genes survive and those without go extinct. Each event has considerable virulence. Attenuation may follow the initial onset of each event, but each of the events starts with substantial virulence.

3. The original asexual species developed some disease-resistant genes. This took a long time because these were new genes that had to evolve.

4. A low-sex-ratio sexually reproducing variation appears. This variation evolved from the asexual species and inherited whatever disease-resistant genes it had. Because of its low sex ratio, the variation is nearly neutral relative to the asexual species in the absence of disease. Also because of the low sex ratio, the males fertilize many females.

5. A highly virulent infectious disease appears and suppresses those without a relatively rare disease-resistant gene. This sets up a strong

competition between the asexual and sexual populations whereby whichever propagates the rare disease-resistant gene the fastest wins. The males fertilize many females and both male and female offspring have disease resistance. This makes more female disease-resistant offspring than the corresponding asexual population so the sexual population with disease resistance grows much faster. Per the Selection Theorem, the differential growth with competition for limited resources causes selection and the sexual variation wins.

6. Virulent infectious diseases appear again and again challenging different and rare disease-resistant genes. Each time a variation with a slightly larger sex ratio competes against the older one. Again a race ensues as to which one can make the most female disease-resistant offspring. The slightly-larger-sex-ratio variations win because they can fertilize more females to produce more female disease-resistant offspring. This is an arms race scenario. A Red Queen variant of the arms race may appear and accelerate evolution, but there are many different evolutionary trajectories that would result in the evolution of sexual reproduction.

7. Eventually, a 50:50 sex-ratio population appears and is stable because an imbalance between the numbers of males and females would be restored to a balance by Fisher's argument.

8. If infectious diseases cease, an invasive asexual species may appear and suppress the sexual species. But this can only happen if the asexual species can successfully compete against the sexual species.

9. Eventually, sexual species will develop sufficient genetic diversity so that it becomes easier to fend off infectious diseases and invasive asexual species.

10. Different infectious-disease sequences will result in different evolutionary trajectories. As a consequence, sexual diversity will appear due to the different trajectories.

The evolution of ideas on the emergence of sexual reproduction has a long history with many important contributions from analyses, fieldwork, experiments, and simulations (see, e.g., Ridley, 1993; and Lively, 2010). The model presented in this chapter builds upon this history and adds a few more elements like replicator modeling of increasing sex ratios. The key feature of the present theory is the emphasis on relatively gradual adaptation for the evolution of sexual reproduction. Gradual adaptation is a hallmark

of evolutionary theory and must also apply to the evolution of sexual reproduction. The means proposed in the present model is the employment of
low-sex-ratio intermediate variations. But any model must stand the test
of empirical data and it is hoped that this model will be considered worthy
of such tests.

7.6 COMMENTS AND CONCLUSIONS

Infectious disease is one of the most powerful evolutionary forces and the
models presented can reflect many of the effects of its key aspects such as
virulence, attenuation, resistance, and coevolution. So what information
have we gained in this chapter? In the first section, it was pointed out that
species with low birth rates are at great risk of extinction from infectious
diseases because even low mortality would be a great danger. For example,
by Eq. 7.2, a species with a replication ratio of 1.1 could be wiped out by an
infectious disease with a mortality of 10%.

However, there is a ray of hope. Living species with very low birth rates
must have very strong immune systems. African forest elephants are an
example of this phenomenon. Their annual birth rate is only about 4.3%
with a consequent doubling time without human impacts of 41 years
(Turkalo et al., 2016). So, what must be special about the elephant immune
system?

A clue comes from the cancer rates in elephants. Because they are large
animals, it might be expected that they might get cancer more often than
smaller animals. But they don't (Caulin and Maley, 2011). There is a gene
called p53 which is vital to stopping mutant cells from becoming can-
cerous. Humans have one copy. Elephants have 20 (Abegglen et al., 2015).
Elephants also have a unique gene called LIF6 that is switched on by p53.
LIF6 causes seeking and destruction in response to DNA damage (Vazquez
et al., 2018). Clearly, more research about other elephant immune-system
advantages could be of considerable value.

In Chapter 2, we wondered why human juvenile mortality was about
50% throughout history up to the twentieth century and did not show any
evolutionary effects. A likely culprit is the infectious-disease phenomenon
known as reservoirs. An infectious disease can exist in one host species
as a reservoir and then frequently infect other host species. The dynamics
will be driven by the reservoir host species and not the secondary host.
This clearly was a major factor in human juvenile mortality because when
numerous health initiatives like clean water and better sanitation were

made in the twentieth century, juvenile mortality was drastically reduced. Obviously, eliminating infectious-disease reservoirs can greatly alter disease mortality. This phenomenon plays an important role in the events described in the next chapter.

Finally, we have described a path by which sexual reproduction could evolve. The driver is the selection pressure from infectious diseases. Gradual adaptation starts with low-sex-ratio species outcompeting asexuals and then competing with each other to increase the sex ratio to 50:50. The competition is dominated by the selection advantage of rare genes that can be spread much faster by males and multiple matings. The faster growth produces selection via the Selection Theorem. And different evolutionary trajectories can create sexual diversity.

If this theory is correct, then the implications are considerable. If males are the bringers of life by bringing disease-resistant genes, the females would have more surviving offspring by choosing them. A male with a disease-resistant gene would tend to be healthier and so females selecting healthy males for mating would benefit offspring. And thus the phenomenon of female preference (e. g., peacock tails and bower birds) is born.

CHAPTER 8

Serengeti

IMAGINE THE GREAT CITY of Chicago with its population of over 2.5 million and its streets, highways, and trains with rivers of humanity on the move. Expand the city limits north to Wisconsin, west to Rockford, and south to Joliet. Now replace the people with wildebeest, gazelles, lions, giraffes, zebras, water buffalo, elephants, and more. Replace the buildings with forests, swamps, rocky hills, grasslands, and woodlands. Replace the streets, highways, and trains with the largest mammal migration on Earth. This is the Serengeti, the great metropolis of megafauna and one of the wonders of the world.

Perhaps because of its beauty and uniqueness, the Serengeti is one of the most studied ecosystems in the world. Few ecosystem studies can match the number of scientific papers and books published about the Serengeti ecosystem. This turned out to be fortuitous because starting in the 1960s, the Serengeti ecosystem experienced a large trophic cascade event that upgraded the entire system. Detailed observation of this upgrading enabled the determination of which species affected others and by how much, with a number of surprises (Sinclair and Norton-Griffiths, 1979). As noted before in Chapter 6, trophic cascades are evolution in action and the Serengeti upgrading was evolution in action on a large scale. Researchers

DOI: 10.1201/9781003391395-8

realized that what they had observed was the detailed rules of life. These rules became known as the Serengeti Rules (Carroll, 2016; Brown, 2018).

In this chapter, we first discuss a few regulating factors and then present an ecosystem model for the upgrading event. We then discuss how the models of evolution and ecological change developed in much of this book are a representation of the Serengeti Rules.

8.1 MIGRATION AND RESIDENCY

Earth is a planet on the move. From tiny microbes to giant whales, evolution has favored the ability to move. The reason is simple—when resources run out, survival is enhanced by the ability to move to regions with more resources.

One of the most dramatic of all movements is the phenomenon of mass migrations. The winner for shear numbers and biomass is the ocean zooplankton that rise from the depths at dusk to feed on phytoplankton and then return at dawn. The marathon long-distance award goes to the Arctic tern that flies from the South Pole before Antarctic winter sets in, zigzags across the Atlantic Ocean with stops in Africa and South America, and spends summers in the Arctic before returning when Arctic winter arrives—an annual round trip of about 40,000 miles. Among megafauna, a contender for the greatest number is the reindeer of the Siberian Taimyr peninsula, which at one time grew to over a million individuals during the annual migration.

But the current record for the greatest megafauna migration is held by the wildebeest of the Serengeti. It starts in the southeast grasslands towards the end of the rainy season and moves northwesterly along the western Serengeti ecosystem. It comprises about 1.2 million wildebeest that are accompanied by up to 600,000 zebras, gazelles, and assorted others. The route varies but the herd has the uncanny ability to find nutritious grass to feed on. The pace varies with numerous stops and starts but averages about 2 to 4 miles per day. The migration reaches the northern Serengeti ecosystem and returns along an eastern route in time for the start of the rainy season in the southeast.

The evolution of migration must have followed as similar a path as the evolution of movement consistent with Selection Model 1.3. If the appearance of a variation for movement results in access to a new resource, then the variation will grow. The same is true for migration. When local resources run out, but migration can provide more resources, then evolution will favor migration (for an extensive discussion of migration see, e.g., Milner-Gulland et al., 2011).

Not all animals migrate. Instead, they evolve residency that keeps them in a local region. For example, the Serengeti crocodiles stay in their river Mara. Obviously, this was driven by less availability of resources outside of the local region. So, migration would not be favored by evolution under this condition.

But there is a third condition for some species. And that is that some migrate and some do not. The species shows both migration and residency at the same time. For all of their great migration, there are some wildebeest that stay home (Estes, 2014). There is less predation for migration than residency (Sinclair et al., 2003), but Chapter 5 showed that this is generally selection neutral for prey. So why is there the evolution of both migration and residency?

Consider the food web in Figure 8.1a and the formulas in Figure 8.1b. The model is Coexistence Model 1.4 but viewed from the perspective of migration. Creature 2 is relatively stationary and consumes the relatively small resource R2. Creature 3 appears that is a variation that migrates outside of the resource R2 area. Migration creates effectively a very much larger resource R3 for creature 3 and resource R3 is the preferred resource for creature 3. Creature 3 population grows while creature 2 drops only slightly because creature 3 prefers the resource available to migration over the resource available to residency. This model demonstrates that migration and residency can coexist. As before in Selection Model 1.3 and Coexistence Model 1.4, the key factor between evolutionary selection of migration and coexistence between migration and residency is resource preference.

8.2 GRASS REGULATION

Evolution is not just about winners and losers. As pointed out in Chapter 1, evolution is also about coexistence. But that does not mean that the coexistence is static and unchanging. Coexistence can also mean regulation—one set of organisms can regulate the abundance of another without leading to extinction. The migration and residency of the wildebeest is but one example. The long and short grasses of the Serengeti are another.

Short grasses dominate the southwestern Serengeti during the winter rainy season. But the western and northern areas have coexisting long and short grasses. The short grass is more nutritious, faster growing, and

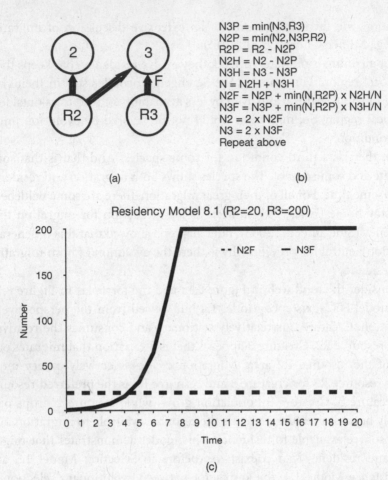

FIGURE 8.1 Residency Model 8.1. (a) The food web is an adaptation of Coexistence Model 1.4. Creature 2 is stationary while creature 3 is migratory. (b) The formulas are also an adaptation of Model 1.4 only this time resource R3 representing resources for migration are much larger. (c) The results demonstrate the coexistence of migration with residency.

has adapted to intense grazing. The long grass is a superior light competitor but is susceptible to grazing. So the short grass can bloom when the long grass is suppressed by the wildebeest. It appears that evolution has maintained coexistence for both types even though the grazing-free long grass can suppress the short grass by shading it from sunlight (Hartvigsen and McNaughton, 1995).

Sun-shading of lower-story plants is a common phenomenon. Consider the food web shown in Figure 8.2a.

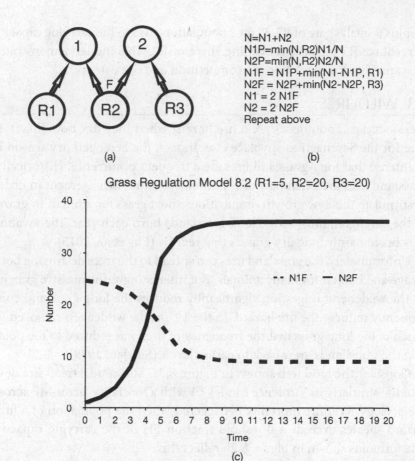

N=N1+N2
N1P=min(N,R2)N1/N
N2P=min(N,R2)N2/N
N1F = N1P+min(N1-N1P, R1)
N2F = N2P+min(N2-N2P, R3)
N1 = 2 N1F
N2 = 2 N2F
Repeat above

(a)   (b)

Grass Regulation Model 8.2 (R1=5, R2=20, R3=20)

(c)

FIGURE 8.2   Grass Regulation Model 8.2. (a) The food web shows a shared resource R2 (sunlight) supplemented by additional resources R1 (shaded light) and R3 (water from deep roots). (b) The formulas show the usual sharing and unfed feeding. (c) The results show that competition and coexistence can lead to regulation rather than just extinction.

Resource R2 (e.g., sunlight) is a shared resource for plants 1 and 2. Plant 2 has an additional significant resource R3 (e.g., water from deeper roots). Plant 1 has a small additional resource (e,g., shaded light). Resource R2 is consumed first. The formulas in Figure 8.2b reflect the expected computation of sharing a resource followed by totaling with the consumption by unfed plants of other resources.

The results are shown in Figure 8.2c. Plant 1 consumes resources R1 and R2. Plant 2 is an invasive specie that consumes resource R3 and shared resource R2. Plant 1 population decreases to carrying capacity of resource

R1 plus a small share of R2. Plant 2 population rises to the carrying capacity of resource R3 plus the remaining share of R2. This model demonstrates the capability for regulation via competition and coexistence.

8.3 WILDFIRES

Trees, shrubs, and grasses are a fire hazard when they dry out. This is as true for the Serengeti as anyplace else. In fact, the Serengeti dry season is so intense that long-grassland fires are a frequent occurrence. Historically, grassland fires are often part of traditional savanna management in order to stimulate the new growth of nutritious short grass for animals to graze. In the Serengeti, most of the long grasslands burn each year. The savanna fires burn mainly long dry grasses that regrow (Eby et al., 2015).

Unfortunately, the grassland fires can spread to the trees destroying both forage and habitat for many animals. But interestingly, the massive grazing by the wildebeest migration significantly reduces the long grass and consequently reduces the fire hazard. In the 1970s, the wildebeest removed so much of the long grass that the frequency of fires was reduced to the point that the woodlands grew back to earlier levels (Sinclair, 1979).

Consider the food web shown in Figure 8.3a. This model treats fire devastation similarly to Virulence Model 7.1 with a mortality factor MF across the ecology. Wildfire mortality is triggered when the population of a fire-hazard species exceeds a threshold fraction FH of the carrying capacity. The formulas shown in Figure 8.2b reflect this.

The results are shown in Figure 8.3c. Plant 1 starts at equilibrium consuming resource R1. Fire-hazard plant 2 appears as an invasive specie and grows rapidly. Plant 2 population exceeds the fire-hazard threshold and ignites wildfires that devastate the ecology. Afterwards, regrowth occurs and the cycle repeats. The results demonstrate the modeling of key features of wildfire hazard.

8.4 ECOSYSTEM MODEL

In the 1890s, domestic cattle near the Serengeti acquired the devastating pathogen known as rinderpest. Rinderpest is due to a virus that is very similar to measles in humans. The rinderpest was an invasive specie that was carried by cattle imported from India (Sinclair, 2012). Wildebeest were subsequently devastated by rinderpest. In the 1960s, cattle vaccination succeeded in eliminating rinderpest. But rinderpest also subsequently vanished in wildebeest proving cattle were the reservoir for the pathogen (Sinclair, 1977).

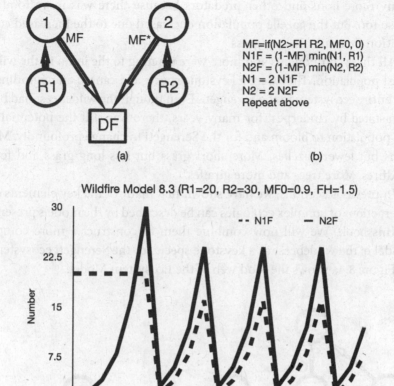

Wildfire Model 8.3 (R1=20, R2=30, MF0=0.9, FH=1.5)

FIGURE 8.3   Wildfire Model 8.3. (a) The food web shows two plants with potential wildfire mortality MF. Plant 2 is an invasive species that can trigger the fire hazard as denoted by MF*. (b). The formulas show that wildfire mortality is triggered by the plant 2 population exceeding a fraction FH of its carrying capacity R2. (c) The results show that overgrowth of plant 2 ignites wildfires that devastate the ecology followed by regrowth and cycle repeat. This demonstrates key features of the wildfire hazard.

Wildebeest populations increased over five times. This blooming of a large mammal population was a complete surprise to everyone and led to a number of follow-on effects. By consuming the long grass, there was less fuel for fire and subsequently fewer wildfires (Holdo et al., 2009). This allowed both the short grass and the tree population to increase. The trees provided more food for giraffes and for many other species. There were many more lions and other predators because there was more food for those too. But the gazelle population decreased due to the increased competition for available long grass.

All these components and more were reacting to the jump in the wildebeest population. They were a keystone specie and caused an upgrading of the entire ecosystem in the Serengeti. Even though the wildebeest had been devastated by rinderpest for many years, there was still the potential for the population to bloom and for the Serengeti to change profoundly. More lions but fewer gazelles. More short grass but less long grass and fewer wildfires. More trees and more giraffes.

In previous sections, we have shown that many of the key elements and interactions of complex ecologies can be described by the models presented in this book. We will now combine them to construct a more complex model of the wildebeest as a keystone specie for the Serengeti ecosystem.

Figure 8.4a shows the food web of the Ecosystem Model.

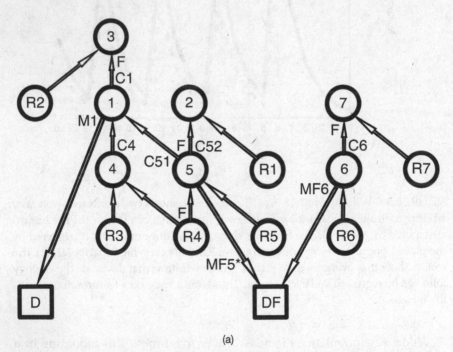

(a)

FIGURE 8.4A   Ecosystem Model Food Web. 1=wildebeest, 2=gazelles, 3=lions, 4= short grass, 5=long grass, 6=trees, 7=giraffes, D=detritus, DF=wildfire detritus, R= resources, M1=pathogen mortality rate, MF=wildfire mortality rate, C=counter-measure constants, F=consume first, *=wildfire starter. R1 through R7 are supplemental resources.

The Ecosystem Model is derived from several previous models: Base Arms Race Model (BARM), Apex Predator 6.4, Virulence 7.1, Grass Regulation 8.2, and Wildfires 8.3. The Ecosystem Model is a representation of a complex ecology with 7 species, 7 resources, predators, prey, pathogens, and wildfires. The model starts with the presence of a pathogen which is then eliminated to observe the effects.

The formulas and the parameters are shown in Figure 8.4b.

Formulas

Parameters

$f(x,y)=min(x,y)/x$
$g(x,y)=max(0,x-y)$ R1=4
MF5=if(N5>FH R4,MF05,0)   R2=5
MF6=if(N5>FH R4,MF06,0)   R3=3
N45=N4+N5   R4=20
N4P=f(N45,R4)N4   R5=10
N4Q=N4P+min(R3,N4-N4P)   R6=20
N5P=f(N45,R4)N5   R7=2
N5Q=N5P+min(R5,N5-N5P)   MF05=0.05
RN5=min(g(N5Q,C51),g(N5Q,C52))   MF06=0.35
N12=N1+N2   FH=1
N1P=f(N12,RN5)N1   C1=6
N2P=f(N12,RN5)N2   C4=13
N5T=N5Q-N1P-N2P   C51=5
N1Q=N1P+min(N1-N1P,g(N5T,C51))   C52=15
N2Q=N2P+min(N2-N2P,g(N5T,C52))   C6=10
N1R=min(N1-N1Q,g(N4Q,C4))
N1S=N1Q+N1R
N3P=min(N3,g(N1S,C1))
N6P=min(N6,R6)
N7P=min(N7,g(N6P,C6))
N1F=(N1S-N3P)(1-M1)
N2F=N2Q+min(N2-N2Q,R1)
N3F=N3P+min(N3-N3P,R2)
N4F=N4Q-N1R
N5F=(N5Q-N1Q-N2Q)(1-MF5)
N6F=(N6P-N7P)(1-MF6)
N7F=N7P+min(R7,N7-N7P)
N1=2 N1F
N2=2 N2F
N3=2 N3F
N4=2 N4F
N5=2 N5F
N6=2 N6F
N7=2 N7F
Repeat above

(b)

FIGURE 8.4B   Ecosystem Model Formulas and Parameters. (1=wildebeest, 2= gazelles, 3=lions, 4=short grass, 5=long grass, 6=trees, 7=giraffes, R= resources, M1=pathogen mortality rate, MF=wildfire mortality rate, C=countermeasure constants, FH=fire-hazard threshold. R1 through R7 are supplemental resources.)

The functions f and g are the usual ones used to simplify formulas. MF5 and MF6 are the wildfire mortalities for long grass 5 and trees 6 respectively that are activated by the long grass 5 population exceeding the fire hazard fraction FH of resource R4. Short grass 4 and long grass 5 share resource R4 and then feed the unfed grass on their exclusive resources R3 and R5 respectively.

Wildebeest 1 and gazelles 2 first share that of long grass 5 which is available to both given the countermeasure constants and then the unfed consume the remainder. Unfed wildebeest 1 get to consume a little more feeding on short grass 4. Lions 3 feed on wildebeest 1. Trees 6 feed on resource R6. Giraffes 7 feed on trees 6.

Finally, the population of surviving wildebeest 1 is adjusted for pathogen mortality. Unfed gazelles 2 feed on resource R1. Unfed lions 3 feed on resource R2. Short grass 4 is computed. Long grass 5 and trees 6 are affected by wildfire mortality. Unfed giraffes 7 feed on resource R7. And replication takes place.

Please note again that in this model, as well as in all models in this book, the choices of parameters and the corresponding number scales are intended for concept demonstration purposes only. Parameter adjustment and parameter fitting would be required for comparison to real data. However, the objective here is not to make detailed ecological models, but rather to explore and illuminate evolutionary processes.

The results are shown in Figure 8.4c–e. The pathogen mortality M1 is changed from 0.5 to 0 at t=10. In Figure 8.4c, the wildebeest 1 population shifts upwards. Gazelle 2 population drops because of increased competition with wildebeest 1 for consumption of long grass 5. Predator lions 3 population shifts upwards because of the increase in prey wildebeest 1.

Figure 8.4d again shows the wildebeest 1 population shifting upwards because of the elimination of pathogen mortality. But it also shows the long grass 5 population decreasing because of increased consumption by wildebeest 1. As a consequence, the short grass 4 population increases because of decreased competition with long grass 5.

Figure 8.4e again shows the long grass 5 population dropping because of increased consumption by wildebeest 1. The frequency of wildfires and corresponding mortality drops because of the drop in long grass 5 population. The trees 6 population shifts upwards because of the drop in wildfire mortality. The giraffe 7 population also shifts upwards because of the increase in consumed trees 6.

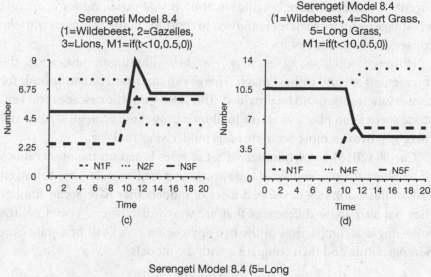

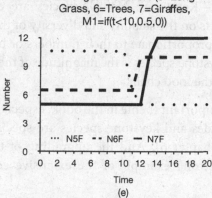

FIGURE 8.4C–E Ecosystem Model Results. The pathogen mortality M1 shifts from 0.5 to 0 at t=10. (c) Wildebeest 1 shift upwards while the gazelles 2 decrease and the lions 3 increase. (d) With the shift of wildebeest 1, long grass 5 decreases and short grass 4 increases. (e) With the shift down of long grass 5 and the drop in wildfires, trees 6 and giraffes 7 increase.

8.5 SERENGETI RULES

In a relatively unchanging ecosystem, the strength of interactions in a food web is difficult to estimate. For example, without change, it can be problematic to know which predators are consuming how much of various prey. Whereas this issue may seem academic, there can be dysfunctional conservation decisions based on poor knowledge like the various historical calls for culling the populations of elephants, wildebeest, and lions in the

Serengeti. An even more horrific situation is that poaching for economic advantage using wire snares is allowed to continue in the Serengeti with the consequent negative results.

However, a changing ecology can help illuminate just what the interactions are and by how much. Thereby an improved scientific basis for conservation action can be provided. This is why trophic cascades and keystone species can play a vital role in illuminating these ecological rules and thereby provide a more accurate basis for decision-making.

Carroll (2016) has synthesized a set of rules based on the observations of trophic cascades and keystone species that he has called the Serengeti Rules. This book has developed a set of models that have some similarities but also some differences that are worth discussion. Therefore, the following is a comparison of the two approaches. We shall first state each Serengeti Rule and then compare it with the models.

> Serengeti Rule 1—Keystones: Not all species are equal. Some species exert effects on the stability and diversity of their communities that are disproportionate to their numbers or biomass. The importance of keystone species is the magnitude of their influence, not their rung in the food chain.

This has been an important theme in the book, especially in Chapters 6 and 8. Trophic cascades and keystone species are specific cases of ecological sensitivity (Chapter 6). And the upgrading of the Serengeti by the wildebeest is a strong example of the extensive ecosystem effects (Chapter 8).

> Serengeti Rule 2—Some species mediate strong indirect effects through trophic cascades. Some members of food webs have disproportionately strong (top-down) effects that ripple through communities and indirectly affect species at lower trophic levels.

Again, this is a theme of both Chapters 6 and 8. The important point is that the models have shown the capability to model trophic cascades whereby species at lower levels can be indirectly affected by top-down effects.

> Serengeti Rule 3—Competition: Some species compete for common resources. Species that compete for space, food, or habitat can regulate the abundance of other species.

It is necessary to expand upon the word "some." If the population of a single species has not hit the carrying capacity, then the members of that species do not have to compete for resources. But they still have to compete for escape from predation, resistance to infectious diseases, and mating for sexual species. However, if the population has hit the carrying capacity, then the members compete for the limited resource. If there are multiple resources, then selection or coexistence between species can result depending on resource preference. With coexistence, some members of the same species can regulate the abundance of other members. The same is true for two species sharing the same resources. These phenomena have been reflected in several models such as that for migration and residency (Section 8.1) and grass regulation (Section 8.2).

> Serengeti Rule 4—Body size affects the mode of regulation. Animal body size is an important determinant of the mechanism of population regulation in food webs, with smaller animals regulated by predators (top-down regulation) and larger animals by food supply (bottom-up regulation).

Chapter 5 makes explicit use of this rule to derive countermeasure functions that lead to the evolution of complexity. The Serengeti Ecosystem Model 8.4 also makes note of this phenomenon by having predation for wildebeest but not for giraffes. The giraffes are regulated by the food supply from trees.

> Serengeti Rule 5—Density: The regulation of some species depends on their density. Some animal populations are regulated by density-dependent factors that tend to stabilize population size.

Again, it is necessary to expand upon the word "some." If predation and infectious diseases are relatively light, then all species with populations below the carrying capacity can experience exponential growth but stop expanding when they hit the carrying capacity. However, if predation or infectious diseases are relatively heavy, they both can limit the population sizes of species. Numerous models in this book reflect this phenomenon.

> Serengeti Rule 6—Migration increases animal numbers. Migration increases animal numbers by increasing access to food (reducing bottom-up regulation) and decreasing susceptibility to predation (reducing top-down regulation).

Section 8.1 on migration and residency discusses this phenomenon in more detail and notes that migration or residency can either be selected or coexist depending on resource preference.

In conclusion, the models of evolution and ecological change developed in much of this book are a representation of the Serengeti Rules. Thus, the models derived from basic elements are in conceptual accord with the observations of natural ecological phenomena.

CHAPTER 9

Summary Discussion

E VOLUTION, ECOLOGICAL CHANGE, AND resource-flow physics are bound together in an eternal golden braid. Evolution causes ecological change and ecological change causes evolution. From the very beginning, creatures altered their environments. At first, there were plenty of flowing resources so their numbers grew exponentially. But then, they hit the resource-limit carrying-capacity resulting in competition and detritus production. The replicators that grew a little faster hogged the resources and pushed out the slowpokes. Thus, evolutionary selection was born. But with multiple resources, resource preferences appeared among the replicators and coexistence was born. Coexistence in turn allowed the appearance of cooperation and even symbiosis.

But the slowpokes had their day too. By becoming detritus, they became the vital fertilizer of new ecologies in which life could flourish. It started with marine snow. Layer after layer, year after year, millennia after millennia, eon after eon, the ocean currents spread the marine snow far and wide and covered the ocean floor. Evaporation, wind, and rain spread the marine snow to land. But changing sea levels left vast areas ripe for the next phases of evolution.

But in the meantime, the engine of variation and selection churned on continuing to make replicators with surviving populations that grew faster.

DOI: 10.1201/9781003391395-9

There was the inevitable tradeoff between streamlining and survival, but the replicator species that produced more surviving offspring was selected over the competition. The board was now set for the next great phase in evolution—the appearance of scavengers.

At first, the population of scavengers bloomed spectacularly given all the detritus that had been created. But once the low-hanging fruit had been picked, the scavenger population crashed and was reduced to whatever the flow of detritus could provide. Scavengers started to differentiate according to the various ages of detritus. Old detritus was less nutritious, but there was much more of it. New detritus was more nutritious, but there was less of it. Eventually, the scavengers evolved into specialists and generalists with the generalists winning if they preferred the specialist resources first. Fertility and scavenging enabled metabolism and the genetic code to grow.

Sometime during the early phases of evolution, sunlight became a new resource and photosynthesis was born. The great oxidation event occurred and the biosphere switched from anaerobic to aerobic. In addition, the great endosymbiotic event occurred and the resulting mitochondria fueled a cornucopia of energy availability. Thus much greater energy became available to power the next phases of evolution.

It was a relatively small evolutionary step to go from a scavenger consuming fresh detritus to a predator consuming prey. But predators had several evolutionary challenges that they had to overcome. For example, overfeeding and cannibalism could cause their extinction. In addition, generalization and prey defense had to evolve. But when the challenges were overcome, predators became a major force for the evolution of complexity and novelty. Coevolutionary arms races ensued that forced predators and prey to constantly evolve better and better offensive and defensive adaptations. Because they regularly brush with close-to-founder effects, predator-prey cycles like that of the Canadian lynx and snowshoe hare seem to act like an ecological echo of these arms races. Whole ecologies were built around coevolution of the myriad of different offensive and defensive tactics. And multicellularity was born. Evolution would change the ecology and the ecological change would alter evolution.

Sometime during the early phases of evolution, parasites and pathogens appeared that hijacked the metabolisms of hosts in order to make more parasites and pathogens. It is possible that the parasites and pathogens predated the scavengers. However, it seems more evolutionarily complicated to hijack live machinery and reprogram it than to just scavenge dead machinery. Nevertheless, the appearance of parasites and pathogens also became a major evolutionary force. Coevolutionary arms races with hosts appeared resulting in endless cycles of virulence, attenuation, and resistance, creating the emergence of complex immune systems. But sexual reproduction also emerged as a result and evolution entered a great new phase of the development of complexity. Sexual species became a storehouse of genetic diversity that could rapidly evolve adaptations to all kinds of selection pressures. And the evolution of complexity and novelty went into overdrive.

Richard Dawkins was right about how replicators can drive evolution. Replicator models and their more complex versions have proven to be useful conceptual tools in illuminating the effects of evolution and ecological change. They have provided insight to many topics of interest from selection to complex ecosystems. The Selection Theorems are a powerful explanation of how selection can occur naturally and quickly. For example, streamlining is a natural consequence of a simple environment and variations that discard unused baggage and thereby grow more quickly.

Preference was well known in evolutionary theory as a strong force in the phenomenon of female or sexual preference. But many organisms also show resource preference. The addition of preferences for resources opens the door for explanations of numerous phenomena. For example, coexistence rather than selection can be a natural consequence. This in turn leads to a better understanding of specialization versus generalization, migration versus residency, regulation, and other phenomena.

Replicator models have provided a foundation for understanding the evolution of complexity. Coevolution can build complexity and novelty, but arms races and Red Queen scenarios can speed it up. Both predator-prey and infectious-disease-host coevolutions can do this and build complexity and novelty. Predator-prey interactions can lead to faster, stronger, and smarter organisms. Infectious-disease interactions with hosts can lead to stronger immune systems and the evolution of sexual reproduction.

Finally, complex ecosystems are a natural consequence of the evolution of complexity and novelty. Our models have provided explanatory power there too. Predator-prey cycles are a natural consequence of shifting resource preference due to changing prey abundance. Keystone species and trophic cascades highlight the rules of ecological interactions. And the Serengeti upgrading is evolution in action on a large scale and shows how deep the interactions can go for a complex ecosystem. Our models are in conceptual agreement with these complex natural phenomena and a representation of the Serengeti Rules.

Sean Carroll was right. The quest to discover how life works matters greatly. One example is the potential value of the very pathogen-resistant immune systems of low-fertility species such as elephants. The value for medical research based on elephant immune systems cannot be overestimated.

A key to evolution is resources: finding new ones while avoiding being found as one. Harold Morowitz was right. Energy and matter resource flow is critical. There is a tyranny of arithmetic: the accounts must balance for resource generation and consumption.

We have now arrived at our goal to illuminate the "why" of evolution and ecological change. So how does complex evolution arise automatically? Evolution and ecologies make an endless series of adaptations. But resources must keep flowing. And the parts must be rich enough with features to combine and form more novel and complex internal and external ecosystems. We found that certain critical features appear to be both necessary and sufficient for enabling complex evolution. These are resources, replication, food webs, detritus, variations, resource preference, mortality, invasive species, predation, kin selection, and countermeasures. As long as those features are present, evolution can grow in an open-ended manner.

So many natural behaviors being derived from our replicator and resource-flow models implies that progress has been made in illuminating the elements of evolution. It is said in mathematics, and physics too, that one is getting somewhere when things become beautiful. And the plains and savannas of the Serengeti are very beautiful.

The quest started by Darwin for the theory of evolution continues. As Darwin (1860) noted:

> There is grandeur in this view of life, with its several powers, having been originally breathed by the Creator into a few forms or into one; and that, whilst this planet has gone cycling on according to the fixed law of gravity, from so simple a beginning endless forms most beautiful and most wonderful have been and are being evolved.

Density-Dependent Regulation

W HEN THE RATE OF population increase is compared to the population size, all populations show an initial increase followed by a drop to zero or below. This is called density-dependent regulation. Any population showing a region of exponential growth will have a linear relationship between rate and size in that region. For

$$N(t)=A\exp(bt), \text{ then}$$

$$dN/dt = bN$$

In addition, all populations are limited by carrying capacities which will cause a drop to zero rate for populations at the respective carrying capacity.

Consider the model shown in Figure A.1a. This model is Simple Replicator Model 1.1 for replication ratio K. Using the formulas in Figure A.1b, compute the population increase:

For $N1(t)<R1$

$$N1F(t)-N1F(t-1)=(K-1)N1F(t-1) \tag{A.1}$$

And for $N1(t)>R1$ and $N1(t-1)>R1$

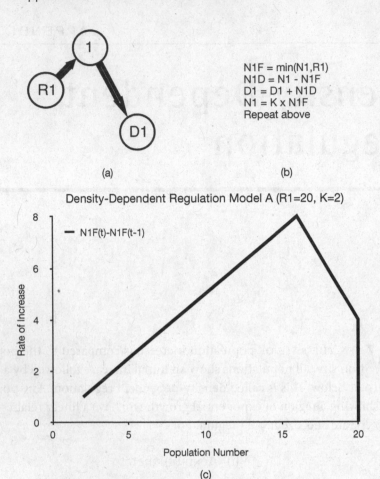

(a)

(b)

N1F = min(N1,R1)
N1D = N1 - N1F
D1 = D1 + N1D
N1 = K x N1F
Repeat above

(c)

FIGURE A.1   Density-Dependent Regulation Model A. (a) Food web for Carrying Capacity Model 1.1. (b) The formulas are generalized from Model 1.1 for replication ratio K. (c) The resulting rate-vs-population graph for K=2.

$$N1F(t)-N1F(t-1)=0 \qquad (A.2)$$

As shown again in Figure A.1c for K=2, this model shows a linear relationship below carrying capacity and zero rate above it. Controlled-carrying-capacity laboratory experiments are predicted to show a sharp peak as shown. However, one must include data below the carrying capacity to show the linear region.

Unfortunately, fluctuations of resources can strongly affect the rate-vs-population graphs. To demonstrate this effect, we create a simulation model that starts with Model A1 and adds normally-distributed random variations of resource R1:

$$R1(t)=R1(t-1)+norminv(rand,0,RS) \qquad (A.3)$$

where RS is standard deviation of a zero-mean additive random number. The frequency spectrum of this simulated resource is similar to many natural processes.

One resulting time-sequence realization for RS=1 is shown in Figure A.2d.

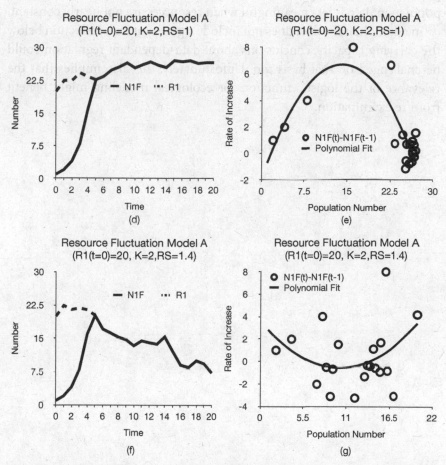

FIGURE A.2 Resource Fluctuation Model A. (d) One time-sequence realization for Model A1 with a simulated time-varying resource R1(t)=R1(t-1)+ norminv(rand,0,RS) with RS=1. (e) Consequent rate-vs-population graph for simulated date in (d). Polynomial fit resembles logistic function. (f) Different time-sequence realization for RS=1.4. (g) Rate-vs-population graph for simulated data from (f). Polynomial fit is not meaningful. This demonstrates that considerable care is needed about drawing conclusions from data about density-dependent regulation.

The population rises to R1 and tracks it afterwards. Figure A.2e shows the consequent rate-vs-population graph. A polynomial fit to the simulation data resembles the logistic function for RS=1.

However, a larger RS=1.4 produces a greater resource R1 variation with one simulation showing the time results in Figure A.2f and rate results in Figure A.2g. Clearly, the rate-vs-population graph is meaningless. The reason is that after the population hits the carrying capacity, the rate-vs-population graph only reflects resource variations. Unfortunately, rate-vs-population data is less meaningful when resources are not nearly constant.

In addition, if the data does not include significant data collection below the carrying capacity, conclusions about data-dependent regulation could be challenged on that basis too. Unfortunately, this also implies that the relevance of the logistic function for ecological modeling might benefit from re-examination.

Selection Theorem Proofs

THIS APPENDIX CONTAINS THE mathematical-proof details of the Selection Theorems.

<u>Selection Theorem:</u> Given two replicator populations competing for a finite resource and the average of random sampling, then the sequence for the population with less growth converges to zero and the sequence for the population with more growth converges to the carrying capacity.

<u>Proof:</u> Let $M(t)$ and $N(t)$ be the populations of competing replicators that have grown to the carrying capacity R at time=t. Thus

$$M(t) + N(t) = R \qquad\qquad (B.1)$$

Let G and H be their corresponding growth factors, respectively. At time=$t+1$, the populations are given by random-sampling averaging as

$$M(t+1) = R\, G\, M(t)/(G\, M(t) + H\, N(t)) \text{ and} \qquad (B.2)$$

$$N(t+1) = R\, H\, M(t)/(G\, M(t) + H\, N(t)) \qquad (B.3)$$

Note that $M(t+1)$ and $N(t+1)$ are also at the carrying capacity such that

$$M(t+1) + N(t+1) = R \tag{B.4}$$

Now compute the population at time=t+2:

$$M(t+2) = R\,G\,M(t+1)/(G\,M(t+1) + H\,N(t+1)) \tag{B.5}$$

Inserting Eq. B.2 we obtain

$$M(t+2) = R\,G^2\,M(t)\,/(G^2\,M(t) + H^2\,N(t)) \tag{B.6}$$

Now compute the incremental population ratio A(t) where

$$A(t) = M(t+2)/M(t+1) \tag{B.7}$$

With some algebra, we obtain

$$A(t) = 1 + (G - H)\,H\,N(t)/(G^2\,M(t) + H^2\,N(t)) \tag{B.8}$$

Let $G < H$, then $A(t) < 1$. Thus, the sequence $\{M(t+n) \mid n=1,2\ldots\}$ is a monotone decreasing sequence (Abbott, 2015, Definition 2.4.1). The sequence $\{M(t+n)\}$ is also bounded between 0 and R. Therefore by the Monotone Convergence Theorem (Abbott, 2015, Theorem 2.4.2) and Eq. B.4, the sequences $\{M(t+n) \mid n=1,2\ldots\}$ and $\{N(t+n) \mid n=1,2\ldots\}$ converge.

Now consider the limit as t->inf and define

$$ML = \lim(t\text{->inf})\ M(t) \tag{B.9}$$

$$NL = \lim(t\text{->inf})\ N(t),\ \text{and} \tag{B.10}$$

$$AL = \lim(t\text{->inf})\ A(t) \tag{B.11}$$

Note that

$$M(t+2) = A(t)\,M(t+1) \tag{B.12}$$

Then by the Algebraic Limit Theorem (Abbot, 2015, Theorem 2.3.3), we have

$$AL = 1 + (G - H) H NL/(G^2 ML + H^2 NL), \qquad (B.13)$$

$$ML + NL = R, \text{ and} \qquad (B.14)$$

$$ML = AL\ ML \qquad (B.15)$$

But since $G < H$, then $AL < 1$. Thus, $ML = 0$ and $NL = R$. QED.

<u>Selection Speed Theorem:</u> The speed of going to zero or the carrying capacity is determined by the magnitude of the growth difference for replicator populations under competition for a limited resource and random sample averaging.

<u>Proof:</u> As shown in Eq. B.8, the incremental population ratio $A=M(t+2)/M(t+1)$ decreases with decreasing growth rate $G<H$. Thus, the sequence $\{M(t+n) \mid n=1,2\ldots\}$ converges to zero more quickly with the larger-growth population sequence $\{N(t+n) \mid n=1,2\ldots\}$ converging to the carrying capacity more quickly. QED.

<u>Extended Selection Theorem:</u> The Selection and Selection Speed Theorems are valid for differential consumption.

<u>Proof:</u> Let U and V be the resource consumption per replicator per unit time of the respective populations for $M(t)$ and $N(t)$. Let the carrying capacity R be such that

$$U M(t) + V N(t) = R \qquad (B.16)$$

Let G and H be the growth factors respectively. At time=t+1 the populations are

$$M(t+1) = R G M(t)/(G U M(t) + H V N(t)) \text{ and} \qquad (B.17)$$

$$N(t+1) = R H N(t)/(G U M(t) + H V N(t)) \qquad (B.18)$$

Compute the population $M(t+2)$ at time=t+2 and the ratio $A(t)=M(t+2)/M(t+1)$ and with algebra it becomes

$$A(t) = 1 + (G - H) H V N(t)/(G^2 U M(t) + H^2 V N(t)) \quad (B.19)$$

If $G < H$, then $A(t) < 1$. The remainder of the proofs follows as before. QED.

References

Abbot, S. (2015) *Understanding Analysis*. New York, NY: Springer.

Abegglen, L. M., A. F. Caulin, A. Chan, K. Lee et al. (2015) "Potential mechanisms for cancer resistance in elephants and comparative cellular response to DNA damage in humans." *Journal of the American Medical Association* 314: 1850–1860.

Alon, U. (2006) *An Introduction to Systems Biology: Design Principles of Biological Circuits*. and 2nd edition (2020). New York, NY: CRC Press.

Alves, J. M., M. Carneiro, J. Y. Cheng, A. L. de Matos et al. (2019) "Parallel adaptation of rabbit populations to myxoma virus." *Science* 363: 1319–1326.

Anderson, R. M., and May, R. M. (1982) "Coevolution of hosts and parasites." *Parasitology* 85: 411–426.

Atlas, R. M., and R. Bartha (1997) *Microbial ecology—fundamentals and applications*. Menlo Park, CA: Benjamin/Cummings Science Publishing.

Ballard, J. (2013) *Black Bears: A Falcon Pocket Guide*. Guilford, CT: Globe Pequot.

Banzhaf, W. and L. Yamamoto (2015) *Artificial Chemistries*. Cambridge, MA: MIT Press.

Bateson, W. (1909) *Mendel's Principles of Heredity*. Cambridge, UK: Cambridge University Press.

Bedau, M. A., J. S. McCaskill, N. H. Packard, S. Rasmussen, C. Adami, D. G. Green, T. Ikegami, K. Kaneko, and T. S. Ray (2000) "Open problems in artificial life." *Artificial Life* 6: 363–376.

Beebe, W., J. Tee-Van, G. Hollister, J. Crane, and O. Barton (1934) *Half Mile Down*. New York, NY: Harcourt.

Bell, G. (1997) *The Basics of Selection*. New York, NY: Chapman & Hall.

Bengtson, S. (2002) "Origins and early evolution of predation." *Paleontological Society Papers* 8: 289–317.

Bengtson S., and Y. Zhao (1992) "Predatorial borings in late Precambrian mineralized exoskeletons." *Science* 257: 367–69.

Bialek, W. (2012) *Biophysics: Searching for Principles*. Princeton, NJ: Princeton University Press.

Blount, Z. D., J. E. Barrick, C. J. Davidson, and R. E. Lenski (2012) "Genomic analysis of a key innovation in an experimental Escherichia coli population." *Nature* 489: 513–518.

Brodie III, E. D. (2010) "Patterns, process, and the parable of the coffeepot incident: Arms races between newts and snakes from landscapes to molecules." In J. B. Losos (ed.), *In the Light of Evolution: Essays from the Laboratory and Field.* Greenwood Village, CO: Roberts and Co.: 93–120.

Bonner, J. T. (2016) "Foreword: The evolution of multicellularity." In Niklas, K. T. and Newman, S. A. (eds.) (2016) *Multicellularity Origins and Evolution.* Cambridge MA: MIT Press.

Brown, N. (dir.) (2018) *The Serengeti Rules.* Chevy Chase, MD: HHMI Tangled Bank Studios and Passion Planet. film.

Carroll, S. B. (2005) *Endless Forms Most Beautiful.* New York, NY: Norton.

Carroll, S. B. (2008) "Evo–devo and an expanding evolutionary synthesis: a genetic theory of morphological evolution." *Cell* 134: 25–36.

Carroll, S. B. (2016) *The Serengeti Rules: The Quest to Discover How Life Works and Why it Matters.* Princeton, NJ: Princeton University Press.

Carson, R. L. (1961) *The Sea Around Us (Rev. ed.).* New York, NY: Oxford University Press.

Caulin A. F., and C. C. Maley (2011) "Peto's paradox: Evolution's prescription for cancer prevention." *Trends in Ecology & Evolution* 26: 175–182.

Charnov, E. L., and S. K. Ernest (2006) "The offspring-size/clutch-size trade-off in mammals." *American Naturalist* 167: 578–582.

Cockell, C. (2018) *The Equations of Life: How physics shapes evolution.* London, UK: Atlantic Books.

Cordain, L., S. B. Eaton, J. B. Miller, N. Mann, and K. Hill (2002) "The paradoxical nature of hunter-gatherer diets: Meat-based, yet non-atherogenic." *European Journal of Clinical Nutrition* 56: 542–552.

Court, S. J., B. Waclaw, and R. J. Allen (2015) "Lower glycolysis carries a higher flux than any biochemically possible alternative." *Nature Communications* 6: 8427.

Darwin, C. (1859) *On the Origin of Species by Means of Natural Selection, or the Preservation of Favoured Races in the Struggle for Life (1st ed.).* London, UK: John Murray; (1860, 2nd ed.).

Dawkins, R. (1976) *The Selfish Gene.* New York, NY: Oxford University Press.

Dawkins, R. (1986) *The Blind Watchmaker: Why the Evidence of Evolution Reveals a Universe Without Design.* New York, NY: Norton.

Dawkins, R. (2009) *The Greatest Show on Earth: The Evidence for Evolution.* London, UK: Bantam.

Dawkins, R., and J. R. Krebs (1979) "Arms races between and within species." *Proceedings of the Royal Society of London. Series B, Biological Sciences* 205 (1161): 489–511.

Diamond, S. J., R. H. Giles, R. L. Kirkpatrick, and G. J. Griffin, (2000) "Hard mast production before and after the chestnut blight." *Southern Journal of Applied Forestry* 24: 196–201.

Dickman, C. R. (1996) "Impact of exotic generalist predators on the native fauna of Australia." *Wildlife Biology* 2: 185–195.

Dobzhansky, T. (1973) "Nothing in biology makes sense except in the light of evolution." *American Biology Teacher* 35 (3): 125–129.

Eby, S., J. Dempewolf, R. M. Holdo, and K. L. Metzger (2015) "Fire in the Serengeti ecosystem: history, drivers, and consequences." In A. R. E. Sinclair, K. L. Metzger, S. A. R. Mduma, and J. M. Fryxell (eds.), *Serengeti IV*, Chicago, IL: University of Chicago Press: 73–104.

Edmunds, M. (1974) *Defence in Animals*. Harlow, UK: Longman.

Einstein, A., and A. Calaprice (2011) *The Ultimate Quotable Einstein*. Princeton, NJ: Princeton University Press.

Eldredge, N., and S. Gould (1972) "Punctuated equilibria: An alternative to phyletic gradualism." In T. J. M. Stopper (ed.), *Models in Paleobiology*, San Francisco, CA: Cooper and Co: 82–115.

Elgar, M. A., and B. J. Crespi (eds.) (1992) *Cannibalism: Ecology and Evolution Among Diverse Taxa*. Oxford, UK: Oxford University Press.

Emelyanov, V. V. (2001) "Evolutionary relationship of Rickettsiae and mitochondria." *FEBS Letters* 501: 11–18.

Emlen, D. J., and C. Zimmer (2020) *Evolution: Making Sense of Life*. New York, NY: Macmillan Learning.

Estes, R. (2014) *The gnu's world : Serengeti wildebeest ecology and life history*. Berkeley and Los Angeles, CA: University of California Press.

Estes J., D. Doak, A. Springer, and T. Williams (2009) "Causes and consequences of marine mammal population declines in southwest Alaska: a food-web perspective." *Philosophical Transactions of the Royal Society B Biological Sciences* 364: 1647–58.

Estes, J. A., C. H. Peterson, and R. S. Steneck (2010) "Some effects of apex predators in higher-latitude coastal oceans." In J. Terborgh and J. A. Estes (eds.), *Trophic Cascades: Predators, Prey, and the Changing Dynamics of Nature*. Washington, DC: Island Press: 37–53.

Estes, J. A., J. Terborgh, J. S. Brashares, M. E. Power et al. (2011) "Trophic downgrading of planet earth." *Science* 333: 301-306.

Fenner, F. (1983) "The Florey lecture, 1983: biological control as exemplified by smallpox eradication and myxomatosis." *Proceedings of the Royal Society of London. Series B, Biological Sciences* 218: 259–285.

Fisher, R. A. (1918) "The correlation between relatives on the supposition of Mendelian inheritance." *Transactions of the Royal Society of Edinburgh* 52 (2): 399–433.

Fisher, R. A. (1930) *The Genetical Theory of Natural Selection*. Oxford, UK: Clarendon Press.

Forgacs, G.and Newman, S. A. (2005) *Biological Physics of the Developing Embryo*. Cambridge, UK: Cambridge University Press.

Frankel, E. A., D. C. Dewey, and C. D. Keating (2014) "Encapsulation of organic materials in protocells." In Kolb, V. (ed.), *Astrobiology: An Evolutionary Approach*. Boca Raton, FL: CRC Press: 217–255.

Ganz, H. H., and D. Ebert (2010) "Benefits of host genetic diversity for resistance to infection depend on parasite diversity." *Ecology* 91: 1263–1268.

Gause, G. F. (1934) *The Struggle For Existence (1st ed.)*. Baltimore, MD: Williams & Wilkins.

Gericke, N. M., and Hagberg, M. (2007) "Definition of historical models of gene function and their relation to students' understanding of genetics." *Science & Education* 16: 849–881.

Giovannoni, S. J., H. J. Tripp, S. Givan, M. Podar et al. (2005) "Genome streamlining in a cosmopolitan oceanic bacterium." *Science* 309: 1242–1245.

Goldford, J. E., H. Hartman, R. Marsland, and D. Segrè (2019) "Environmental boundary conditions for the origin of life converge to an organo-sulfur metabolism." *Nature Ecology & Evolution* 3: 1715–1724.

Gould, S. J. (1989) *Wonderful Life: The Burgess Shale and the Nature of History*. New York, NY: Norton.

Grant, P. R., and B. R. Grant (2006) "Evolution of character displacement in Darwin's finches." *Science* 313: 224–226.

Hague, M. T. J., A. N. Stokes, C. R. Feldman, E. D. Brodie Jr., and E. D. Brodie III (2020) "The geographic mosaic of arms race coevolution is closely matched to prey population structure." *Evolution Letters* 4 (4): 317–332.

Haldane, J. B. S. (1924) "A mathematical theory of natural and artificial selection. Part I." *Proceedings of the Cambridge Philosophical Society* 23: 19–41.

Haldane, J. B. S. (1929) "Origin of life." *The Rationalist Annual* 148, 3–10.

Hamilton, W. D. (1964) "The genetical evolution of social behaviour. I & II." *Journal of Theoretical Biology* 7: 1–52.

Hamilton, W. D. (1971) "Geometry for the selfish herd." *Journal of Theoretical Biology*. 31 (2): 295–311.

Hardin, G. (1960) "The competitive exclusion principle." *Science* 131 (3409): 1292–1297.

Harrison, S. A., Nunes Palmeira, R. N., Halpern, A., Lane, N. (2022) "A biophysical basis for the emergence of the genetic code in protocells." *Biochimica et Biophysica Acta (BBA) - Bioenergetics*. 148597. ISSN 0005-2728. https://doi.org/10.1016/j.bbabio.2022.148597.

Hartvigsen, G., and S. J. McNaughton (1995) "Tradeoff between height and relative growth rate in a dominant grass from the Serengeti ecosystem." *Oecologia* 102: 273–276.

Hayward, M. W., P. Henschel, J. O'Brien, M. Hofmeyr et al. (2006) "Prey preferences of the leopard (*Panthera pardus*)." *Journal of Zoology* 270 (2): 298–313.

Hendry, A. P. (2017) *Eco-evolutionary Dynamics*. Princeton, NJ: Princeton Univ. Press.

Herron, M. D., J. M. Borin, J. C. Boswell, J. Walker, J. et al. (2019) "De novo origins of multicellularity in response to predation." *Scientific Reports* 9 (1): 2328.

Herron, M. D., and Michod, R. E. (2008) "Evolution of complexity in the volvocine algae: Transitions in individuality through Darwin's eye." *Evolution*, 62, 436–451.

Herron, M. D. and Nedelcu, A. M. (2015) "Volvocine algae: From simple to complex multicellularity." In Ruiz-Trillo, I. and Nedelcu, A. M. (eds.) *Evolutionary Transitions to Multicellular Life , Advances in Marine Genomics 2*. NL: Springer Science+Business Media.

Hewitt, G. (1921) *The Conservation of the Wildlife of Canada*. New York, NY: Charles Scribner's Sons.

Hoagstrom, C. (2014) "Predator-prey cycles." In J. E. Duffy (topic ed.), C. J. Cleveland (ed.) *Encyclopedia of Earth*. Washington, DC: National council for Science and the Environment.

Hogan, C. M. (2014) "Overfishing." In S. Draggan (topic ed.), C. J. Cleveland (ed.) *Encyclopedia of Earth*. Washington, DC: National council for Science and the Environment.

Holdo, R. M., A. R. E. Sinclair, A. P. Dobson, K. L. Metzger et al. (2009) "A disease-mediated trophic cascade in the Serengeti and its implications for ecosystem C." *PLOS Biology* 7 (9): e1000210.

Hörnfeldt, B. (1978) "Synchronous population fluctuations in voles, small game, owls, and tularemia in Northern Sweden." *Oecologia* 32 (2): 141–152.

Hudson, R., R. de Graaf, M. S. Rodin, A. Ohno et al. (2020) "CO_2 reduction driven by a pH gradient," *Proceedings of the National Academy of Sciences USA*: 117 (37): 22873–22879.

Huxley, J. (1942) *Evolution: The Modern Synthesis*. London, UK: Allen & Unwin.

Jaouen, K., M. P. Richards, A. Le Cabec, F. Welker et al. (2019) "Exceptionally high δ15N values in collagen single amino acids confirm Neandertals as high-trophic level carnivores." *Proceedings of the National Academy of Sciences USA* 116: 4928–4933.

Johnson, C. R., S. C. Banks, N. S. Barrett, F. Cazassus et al. (2011) "Climate change cascades: Shifts in oceanography, species' ranges and subtidal marine community dynamics in eastern Tasmania." *Journal of Experimental Marine Biology and Ecology* 400:17–32.

Kerr, P. J. (2012) "Myxomatosis in Australia and Europe: a model for emerging infectious diseases." *Antiviral Research* 93: 387–413.

Kerr, P. J., J. Liu, I. Cattadori, E. Ghedin et al. (2015) "Myxoma virus and the Leporipoxviruses: An evolutionary paradigm." *Viruses* 7: 1020–1061.

Kerr, P. J., I. Cattadori, J. Liu, D. Sim et al. (2017) "Next step in the ongoing arms race between myxoma virus and wild rabbits in Australia is a novel disease phenotype." *Proceedings of the National Academy of Sciences USA* 114 (35): 9397–9402.

Kenyon, K. W. (1969) *The Sea Otter in the Eastern Pacific Ocean. North American Fauna, No. 68*. Washington, D.C: U.S. Government Printing Office,

Kimura, M. (1968) "Evolutionary rate at the molecular level." *Nature* 217 (5129): 624–6.

Kimura, M. (1991) "Recent development of the neutral theory viewed from the Wrightian tradition of theoretical population genetics." *Proceedings of the National Academy of Sciences USA* 88: 5969–5973.

King, K. C., L. F. Delph, J. Jokela, and C. M. Lively (2009) "The geographic mosaic of sex and the Red Queen." *Current Biology* 19: 1438–1441.

Krebs, C. J., R. Boonstra, S. Boutin, and A. R. E. Sinclair (2001) "What drives the 10-years cycle of snowshoe hares?" *BioScience* 51(1): 25–35.

Krumhansl, K. A. K., D. D. K. D. Okamoto, A. Rassweiler, M. Novak et al. (2016) "Global patterns of kelp forest change over the past half-century." *Proceedings of the National Academy of Sciences USA* 113: 13785–13790.

Lane, N. (2009) *Life Ascending: The Ten Greatest Inventions of Evolution*. New York, NY: Norton.

Lane, N. (2015) *The Vital Question: Energy, Evolution and the Origin of Complex Life*. New York, NY: Norton.

Larson, S., R. Jameson, M. Etnier, M. Fleming, and P. Bentzen (2002) "Loss of genetic diversity in sea otters (*Enhydra lutris*) associated with the fur trade of the 18th and 19th centuries." *Molecular Ecology* 11: 1899–1903.

Lenski, R. E. (2017) "Experimental evolution and the dynamics of adaptation and genome evolution in microbial populations." *ISME Journal* 11: 2181–2194.

Lively, C. M. (1987) "Evidence from a New Zealand snail for the maintenance of sex by parasitism." *Nature* 328: 519–521.

Lively, C. M., C. Craddock, and R. C. Vrijenhoek. (1990) "Red Queen hypothesis supported by parasitism in sexual and clonal fish." *Nature* 344: 864–866.

Lively, C. M. (2010) "A review of Red Queen models for the persistence of obligate sexual reproduction." *Journal of Heredity* 101: S13–S20.

Malthus, T. R. (1798) *An Essay on the Principle of Population*. London, UK: W. Pickering.

Margulis, L. (1970) *Origin of Eukaryotic Cells*. New Haven, CT: Yale University Press.

Martin, W., and M. J. Russell (2003) "On the origins of cells: a hypothesis for the evolutionary transitions from abiotic geochemistry to chemoautotrophic prokaryotes, and from prokaryotes to nucleated cells." *Philosophical Transactions Royal Society B* 358: 59–83.

Maynard-Smith, J. (1979) "Game theory and the evolution of behavior." *Proceedings of the Royal Society of London. Series B, Biological Sciences* 205: 475–488.

Mayr, E. (1942) *Systematics and the Origin of Species*. New York, NY: Columbia University Press.

McArthur, J. V. (2006) *Microbial Ecology: An Evolutionary Approach*. Boston, MA: Elsevier.

McGlothlin, J. W., M. E. Kobiela, C. R. Feldman, T. A. Castoe et al. (2016) "Historical contingency in a multigene family facilitates adaptive evolution of toxin resistance." *Current Biology* 26: 1616–1621.

Mendel, G. J.(1866) "Versuche uber Pflanzen-Hybriden." *Verhandlungen des naturforschenden Vereines in Brünn* 4: 3–47 (English translation by W. Bateson, 1909).

Milner-Gulland, E. J., J. M. Fryxell and A. R. E. Sinclair (eds.) (2011) *Animal Migration: A Synthesis*. Oxford, UK: Oxford University Press.

Morell, V. (2014) " 'Carnivorous ballet' helps cheetahs coexist with lions." *Science Post*, Apr. 18, 2014.

Morowitz, H. (1968) *Energy Flow in Biology*. Woodbridge, CT: Ox Bow Press.

Morran, L. T., O. G. Schmidt, I. A. Gelarden, R. C. Parris II, and C. M. Lively (2011) "Running with the red queen: host–parasite coevolution selects for biparental sex." *Science* 333: 216–218.

Murugan, A., K. Husain, M. J. Rust, C. Hepler, J. Bass, J. M. Pietsch, P. S. Swain, S. G. Jena, J. E. Toettcher, A. K. Chakraborty, K. G. Sprenger, T Mora, A. M. Walczak, O. Rivoire, S. Wang, K. B. Wood, A. Skanata, E. Kussell, R. Ranganathan, H-Y. Shih, and N. Goldenfeld (2021) *Physical Biology* 18, 041502.

Neveu, M., H-J. Kim, and S. A. Benner (2013) "The 'strong' RNA world hypothesis: Fifty years old." *Astrobiology* 13: 391–403.

Nilsson, D., and S. Pelger (1994) "A pessimistic estimate of the time required for an eye to evolve." *Proceedings of the Royal Society of London. Series B, Biological Sciences* 256: 53–58.

Oparin, A. I. (1924) *Proiskhozhdenie zhizny*. Moscow, USSR: Izd.Moskovhii Rabochil. (translated by Ann Synge as Oparin, A. I. (1957) *The Origin of Life on Earth*. New York, NY: Academic.)

Packard, N., M. A. Bedau, A. Channon, A., T. Ikegami, S. Rasmussen, K. Stanley, and T. Taylor (2019) "An overview of open-ended evolution: Editorial introduction to the open-ended evolution II special issue." *Artificial Life* 25: 93–103.

Paine, R. T. (1966) "Food web complexity and species diversity." *American Naturalist* 100: 65–75.

Paine, R. T. (1969) "A note on trophic complexity and community stability." *American Naturalist* 103: 91–93.

Paine, R. T. (1974) "Intertidal community structure." *Oecologia* 15: 93–120.

Paine, R. T. (1980) "Food webs: Linkage, interaction strength and community infrastructure." *Journal of Animal Ecology* 49 (3): 666–685.

Peng, C., S. L. Haller, M. M. Rahman, G. McFadden, and S. Rothenburg (2016) "Myxoma virus M156 is a specific inhibitor of rabbit PKR but contains a loss-of-function mutation in Australian virus isolates." *Proceedings of the National Academy of Sciences USA* 113: 3855–3860.

Petráková, L., E. Líznarová, S. Pekár, C. R. Haddad et al. (2015) "Discovery of a monophagous true predator, a specialist termite-eating spider (*Araneae: Ammoxenidae*)." *Scientific Reports* 5: 14013.

Pfennig, D. W. (1997) "Kinship and cannibalism." *Bioscience* 47: 667–675.

Pfennig, D. W., P. W. Sherman, and J. P. Collins (1994) "Kin recognition and cannibalism in polyphenic salamanders." *Behavioral Ecology* 5: 225–232.

Phelps, Q. E., S. J. Tripp, K. R. Bales, D. James et al. (2017) "Incorporating basic and applied approaches to evaluate the effects of invasive Asian Carp on native fishes: a necessary first step for integrated pest management." *PLOS One* 12 (9): e0184081.

Philip, G. K., and S. J. Freeland (2011) "Did evolution select a nonrandom 'alphabet' of amino acids?" *Astrobiology* 11: 235–240.

Pigliucci, M., and G. B. Müller (eds.) (2010) *Evolution: The Extended Synthesis.* Cambridge, MA: MIT Press.

Reimche, J., E. D. Brodie Jr., A. N. Stokes, E. J. Ely et al. (2020) "The geographic mosaic in parallel: Matching patterns of newt tetrodotoxin levels and snake resistance in multiple predator–prey pairs." *Journal of Animal Ecology* 89 (7): 1645–1657.

Rice, S. H. (2004) *Evolutionary Theory: Mathematical and Conceptual Foundations.* Sunderland, MA: Sinauer.

Ridley, M. (1993) *The Red Queen: Sex and the Evolution of Human Nature.* London, UK: Viking.

Ridley, M. (2004) *Evolution.* Malden, MA: Blackwell.

Rokas, A. (2008) "The origins of multicellularity and the early history of the genetic toolkit for animal development." *Annual Review of Genetics* 42:235–251.

Sagan (née Margulis), L. (1967) "On the origin of mitosing cells". *Journal of Theoretical Biology.* 14 (3): 255–274.

Sanders, J. A., and F. Verhulst (1985) *Averaging Methods in Nonlinear Dynamical Systems.* New York, NY: Springer-Verlag.

Schutt, B. (2017) *Cannibalism: A Perfectly Natural History.* Chapel Hill, NC: Algonquin Books.

Sharpe, S. C., Eme, L., Brown, M. W., and Roger, A. J. (2015) "Timing the origins of multicellular eukaryotes through phylogenomics and relaxed molecular clock analyses." In Ruiz-Trillo, I. and Nedelcu, A. M. (eds.) *Evolutionary*

Transitions to Multicellular Life, Advances in Marine Genomics 2. NL: Springer Science+Business Media.

Silver, M. (2015) "Marine snow: A brief historical sketch." *Limnology and Oceanography Bulletin* 24: 5–10.

Sinclair, A. R. E. (1977) *The African Buffalo: A Study of Resource Limitation of Populations*. Chicago, IL: University of Chicago Press.

Sinclair, A. R. E. (1979) "Dynamics of the Serengeti ecosystem." In A. R. E. Sinclair and M. Norton-Griffiths (eds.) *Serengeti: Dynamics of an Ecosystem*, Chicago, IL: University of Chicago Press: 1–30.

Sinclair, A. R. E. (2012) *Serengeti Story*. Oxford, UK: Oxford University Press.

Sinclair, A. R. E. and M. Norton-Griffiths (eds.) (1979) *Serengeti, Dynamics of an Ecosystem*. Chicago, IL: University of Chicago Press.

Sinclair, A. R. E., S. Mduma, and J. S. Brashares (2003) "Patterns of Predation in a Diverse Predator-Prey System." *Nature* 425: 288–290.

Smith, K. F., D. F. Sax, and K. D. Lafferty (2006) "Evidence for the role of infectious disease in species extinction and endangerment." *Conservation Biology* 20: 1349–1357.

Smith, T. M., and R. L. Smith (2006) *Elements of Ecology*. San Francisco, CA: Pearson Benjamin Cummings.

Smuts, G. L., G. A. Robinson, and I. J. Whyte, (1980) "Comparative growth of wild male and female lions (*Panthera leo*)." *Journal of Zoology* 190 (3): 365–373.

Soetaert, K, and P. M. J. Herman (2009) *A Practical Guide to Ecological Modelling. Using R as a Simulation Platform*. New York, NY: Springer-Verlag.

Solari, C. A, Galzenati, V. J., Kessler, J. O. (2015) "The Evolutionary Ecology of Multicellularity: The Volvocine Green Algae as a Case Study." In Ruiz-Trillo, I. and Nedelcu, A. M. (eds.) *Evolutionary Transitions to Multicellular Life, Advances in Marine Genomics 2*. NL: Springer Science+Business Media.

Standish, R. (2003) "Open-ended artificial evolution." *International Journal of Computational Intelligence and Applications* 3: 167–175.

Stupka, A. (1960) *Great Smoky Mountains National Park. Natural History Handbook No. 5*. Washington, DC: U.S. Government Printing Office.

Tokuriki, N., and D. S. Tawfik (2009) "Stability effects of mutations and protein evolvability." *Current Opinion in Structural Biology* 19: 596–604.

Turkalo, A. K., P. H. Wrege, and G. Wittemyer (2016) "Slow intrinsic growth rate in forest elephants indicates recovery from poaching will require decades." *Journal of Applied Ecology* 54 (1): 153–159.

Turner, C. B., Z. D. Blount, D. H. Mitchell, and R. E. Lenski (2015) "Evolution and coexistence in response to a key innovation in a long-term evolution experiment with *Escherichia coli*." bioRxiv 10.1101/020958.

Van Blaricom, G. R., and J. A. Estes (eds.) (1988) *The Community Ecology of Sea Otters (Ecological Studies No. 65)*. New York, NY: Springer-Verlag.

Van Driesche, J., and R. Van Driesche (2000) *Nature Out of Place: Biological Invasions in the Global Age*. Washington, DC: Island Press.

Van Valen, L. (1973) "A new evolutionary law." *Evolutionary Theory* 1: 1–30.

Vazquez, J. M., M. Sulak, S. Chigurupati, and V. J. Lynch (2018) "A Zombie LIF Gene in elephants is upregulated by TP53 to induce apoptosis in response to DNA damage." *Cell Reports* 24: 1765–1776.

Volk, A. A., and J. A. Atkinson (2013) "Infant and child death in the human environment of evolutionary adaptation." *Evolution and Human Behavior* 34: 182–192.

Wagner, A. (2014) *Arrival of the Fittest*. New York, NY: Current.

Watson, J. D., and F. H. Crick (1953) "Molecular structure of nucleic acids; a structure for deoxyribose nucleic acid." *Nature* 171: 737–738.

Wikipedia (2021) "*E. coli* long-term evolution experiment." Last modified 7 October 2021. https://en.wikipedia.org/wiki/E._coli_long-term_evolution_experiment.

Williams, T. M., J. A. Estes, D. F. Doak, and A. M. Springer (2004) "Killer appetites: assessing the role of predators in ecological communities." *Ecology* 85 (12): 3373–3384.

Wiser, M. J., N. Ribeck, and R. E. Lenski (2013) "Long-Term Dynamics of Adaptation in Asexual Populations." *Science* 342 (6164): 1364–1367.

Wood, R. A. (2019) "The rise of animals." *Scientific American* 320 (6): 24–31.

Wright, S. (1931) "Evolution in Mendelian populations." *Genetics* 16: 97–159.

Zachar, I., Á. Kun, C. Fernando, and E. Szathmáry (2010) "Replicators. From molecules to organism." In S. Kernbach (ed.) *Handbook of Collective Robotics: Fundamentals and Challenges*. Boca Raton, FL: CRC Press: 473–501.

Index

Printed in the United States
by Baker & Taylor Publisher Services